University LECTURE Series

Volume 61

Lectures on the Theory of Pure Motives

Jacob P. Murre
Jan Nagel
Chris A. M. Peters

American Mathematical Society
Providence, Rhode Island

2010 *Mathematics Subject Classification.* Primary 14-02, 14C15, 14C25, 19E15.

For additional information and updates on this book, visit
www.ams.org/bookpages/ulect-61

Library of Congress Cataloging-in-Publication Data

Library of Congress Cataloging-in-Publication Data has been applied for by the AMS. See
www.loc.gov/publish/cip/.

Contents

Introduction

The *theory of motives* was created by Grothendieck in the mid-sixties, starting from around 1964. In a letter to Serre dated August, 16 1964 Grothendieck mentioned for the first time the notion of "motives"; see [**Col-Se**, pages 173 and 275]. At that time this notion was still rather vague for him, but he was already beginning to see a precise "yoga" for such a theory (see the letter cited above) and his "motivation"(!) (or at least one of his motivations) must have been the following.

In the early sixties Grothendieck, with the help of Artin and Verdier, had developed étale cohomology theory. From that moment on there existed a cohomology theory for every prime number ℓ different from the characteristic p of the underlying field. Moreover, in characteristic zero there exist also the classical Betti and de Rham theory and for positive characteristic Grothendieck already had the outline for the crystalline cohomology theory.

Hence there was an abundance of "good" (so-called Weil) cohomology theories! But all these theories have similar properties, and in characteristic zero there even are comparison theorems between them: the famous de Rham isomorphism theorem between Betti and de Rham theory, and the Artin isomorphism between Betti and étale cohomology.

There should be a deeper reason behind this! In order to explain and understand this, Grothendieck envisioned a "universal" cohomology theory for algebraic varieties: the *theory of motives*. Grothendieck expected that there should exist a suitable $\mathbb{Q}$-linear semisimple abelian tensor category with "*realization*" functors to all Weil cohomology theories.[1]

The best way to see what Grothendieck had in mind is to quote his own words. In section 16 (les motifs - ou la cœur dans la cœur) of the "En guise d'avant propos" of his "Récoltes et Semailles" [**Groth85**], Grothendieck writes the following:

... Contrairement à ce qui se passait en topologie ordinaire, on se trouve donc placé là devant une abondance déconcertante de théories cohomologiques différentes. On avait l'impression très nette qu'en un sens, qui restait d'abord très flou, toutes ces théories devaient 'revenir au même', qu'elles 'donnaient les mêmes résultats'. C'est pour parvenir à exprimer cette intuiton de 'parenté' entre théories cohomologiques différentes que j'ai dégagé la notion de 'motif' associé à une variété algébrique. Par ce terme j'entends suggérer qu'il s'agit du 'motif commun' (ou de la raison commune) sous-jacent à cette multitude d'invariants cohomologiques différents associés à la variété, à l'aide de la multitude de toutes les théories cohomologiques possibles à priori. [...] . Ainsi, le motif associé à une variété algébrique constituerait

---

[1](M.) I remember that during a private conversation in October or November 1964 Grothendieck told me that he was now developing a new theory that would finally explain the (similar) behaviour of all the different cohomology theories.

l'invariant cohomologique 'ultime', 'par excellence', dont tous les autres (associés aux différentes théories cohomologiques possibles) se déduiraient, comme autant d''incarnations' musicales, ou de 'réalisations' différentes [...].

For nice "elementary" introductions to the ideas and the concept of motives we recommend the papers of Serre [**Serre91**], Mazur [**Mazur**] and the recent preprint of Milne [**Mil09**].

Grothendieck has constructed what we now call the category of *"pure motives"*; these are objects constructed from smooth, projective varieties by means of the theory of algebraic cycles modulo a suitable equivalence relation (roughly speaking motives are a kind of direct summands of algebraic varieties). The equivalence relation Grothendieck had in mind was numerical equivalence (although his construction works for every good equivalence relation). According to him [**Groth69a**, page 198]:

the theory of motives is a systematic theory of the "arithmetic properties" of algebraic varieties as embodied in their group of classes for numerical equivalence

Grothendieck was well aware that finally one needs a more general theory; see his letter to Illusie, reproduced in the appendix of [**Jann94**, p. 296]. He envisaged a category of *"mixed motives"* attached to the category of all (i.e., arbitrary) varieties defined over a field k, in the same way as in the complex case one needs not only Hodge structures for smooth compact varieties but also mixed Hodge structures for arbitrary complex varieties.

Such a theory of "mixed motives" has, up to now, not yet been constructed in a satisfactory way, although important progress has been made on the one hand by the construction of triangulated categories of motives by (independently) Hanamura [**Hana95, Hana04**], Levine [**Lev98**] and Voevodsky [**Maz-Vo-We, Voe00**], and on the other hand by Nori [**Nori**], who constructed a very original candidate for a category of mixed motives (cf. also [**Lev05**, p. 462] and [**Hub-MüS**, Appendix]).

Returning to the category of pure motives: its *construction* is (contrary to some widespread misunderstanding!) entirely *unconditional* and in fact – except for its originality – surprisingly simple! However the question whether these motives have the *required good properties* depends on *conjectures* for algebraic cycles; these are the famous *"standard conjectures"* of Grothendieck formulated in [**Groth69a**], and discussed also in the two papers of Kleiman [**Klei68**], [**Klei94**]. Partially these conjectures center around very deep existence problems for algebraic cycles.

Although up to now very little progress has been made on these conjectures (with one exception: a beautiful result by Jannsen [**Jann92**], see Lecture 3), the *"yoga of motives"* has proved to be a very powerful and influential tool in the development of algebraic geometry and number theory, especially in questions of arithmetical algebraic geometry.

The influence of this yoga is nowadays formidable! To give only some examples:

— in his 1970 lecture [**Del70**] at the ICM congress in Nice, where he outlined his famous work on Hodge theory, Deligne already mentions the inspiration from Grothendieck's theory of motives; see also [**Groth69b**];

— in the papers of Deligne, Milne, Ogus and Shih in [**Del-Mi-Og-Sh**] the influence of this yoga is evident;

> — in 1991 the AMS organized a three week conference on motives, resulting in the appearance of two thick volumes of proceedings full of results and ideas originating from motives [**Jann-Kl-Se**];
>
> — the ideas behind the celebrated work of Voevodsky clearly stem from the yoga of motives.

In fact we could make a very long list! [2]

Grothendieck himself has – unfortunately – published only one paper on motives, namely [**Groth69a**]. But in 1967 he has given a series of lectures on his theory at the IHES and – fortunately – some of the attendants of these lectures have reported on this. In fact, these lectures are at the origin of the well-known paper by Manin [**Manin**], of Demazure's Bourbaki lecture [**Dem**] and also of Kleiman's paper [**Klei70**].[3] [4] [5]

This book deals mainly with pure motives, except for the last two chapters where we have tried to give at least some ideas on relative motives and on further developments in the direction of mixed motives. Also we concentrate only on the geometric aspects of the theory. For the arithmetic aspects we suggest the beautiful book by André [**Andr**].

Grothendieck was aiming for a theory built on *numerical equivalence* of algebraic cycles because, as he did foresee and as was later proved by Jannsen, this – and only this – gives an abelian semisimple category. In the present book we mainly work with motives modulo *rational equivalence*, so-called *"Chow motives"*, the reason being that such motives not only yield information on the cohomology of the underlying variety but also on the Chow groups themselves.

The structure of the book is as follows. We start with a short outline of algebraic cycles and Chow groups because, as we mentioned earlier, these are at the basis of the entire theory. To chapter 1 we have added two appendices; in the first one we give a short survey of the main known results for divisors and algebraic cycles of higher codimension, and in the second one we discuss a theorem on the relation between algebraic and (smash) nilpotent equivalence obtained independently by Voisin and Voevodsky (for the definition of these notions, see chapter 1). Then in chapter 2 we describe Grothendieck's construction of the category of pure motives and some examples, in particular motives of curves. Here we see a striking fact: the category of motives contains as a full subcategory the category of abelian varieties up to isogeny!

In chapter 3 we discuss the standard conjectures followed by the treatment of the celebrated theorem of Jannsen, which shows that the category of motives modulo numerical equivalence is an abelian semisimple category. In chapters 4 and 5 we discuss the remarkable concept of "finite-dimensionality" for motives,

---

[2] in the introduction of "Récoltes et Semailles" [**Groth85**, Introduction (II), p. xviii] Grothendieck states: "Et le "yoga des motifs" auquel m'a conduit cette réalité longtemps ignorée est peut-être le plus puissant instrument de découverte que j'aie dégagé dans cette première période de ma vie de mathématicien."

[3] Manin [**Manin**, p. 444] states: "I learned from the theory of motives from lectures that Grothendieck gave at IHES in the spring of 1967 ... This paper is no more than the fruit of assiduous meditation on these beautiful ideas"

[4] See also footnote 1 in Serre's paper [**Serre91**]

[5] There are also unpublished manuscripts of Grothendieck with further ideas on motives; which –as we have been informed – have a rich content.

due independently to S.-I. Kimura and O'Sullivan. This idea is undoubtedly very important. Kimura and O'Sullivan conjectured that every (Chow) motive is finite dimensional, but for the moment this has only been proved for varieties that are dominated by products of algebraic curves (for the precise statement see chapters 4 and 5); this class contains in particular all abelian varieties. In chapter 4 we treat the definition of finite-dimensionality and the case of an algebraic curve. (The finite-dimensionality of the motive of a curve basically goes back to a theorem of Šermenev in 1971, but to avoid misunderstanding: Šermenev did not have the idea of finite-dimensionality.) In chapter 5 we then discuss the surprising properties resulting from finite-dimensionality; results mainly due to Kimura and O'Sullivan. These results are obtained using representation theory of the symmetric group (partitions, Young diagrams etc.). The next chapter, chapter 6, is on the construction of the so-called Picard and Albanese motives, the distribution of the Chow groups over these motives and on the motive of an algebraic surface. The results in chapter 6 are true unconditionally, but in chapter 7 we discuss a set of conjectures on algebraic cycles supplementing, but mainly independent of, the Standard Conjectures. We first state the Bloch-Beilinson conjectures on the filtration on Chow groups, and next the related conjectures by the first named author. We discuss a theorem of Jannsen saying that these two sets of conjectures are equivalent, and we discuss examples of varieties for which these conjectures – or part of them – can be proved unconditionally, in particular we discuss the results in the case of the product of two surfaces. (This is partly based on joint work of the first author with Kahn and Pedrini.)

After these chapters, all on pure motives, we have added two chapters on further developments. Chapter 8 deals with relative motives; here correspondences should be replaced by relative correspondences, cohomology groups should be replaced by (perverse) direct image sheaves, and the Künneth formula should be replaced by the famous decomposition theorem of Beilinson, Bernstein, Deligne and Gabber. Finally, in chapter 9 we report briefly on the developments in the direction of mixed motives. First we present a construction due to Bittner (and Looijenga) of the so-called "motivic Euler characteristic", then we discuss the "motivic weight complex" of Gillet-Soulé (see also [**Gu-Na**]), and finally we give a very short presentation of the construction of Voevodsky's triangulated category of mixed motives. These last two chapters are only intended to give some idea of these concepts and developments; for a thorough treatment of the material in these chapters we advise the reader to consult the original paper of Corti-Hanamura [**Cor-Ha00**] for relative motives and the books of André [**Andr**], Levine [**Lev98**] and Mazza-Voevodsky-Weibel [**Maz-Vo-We**] for mixed motives.

Acknowledgment.
We would like to thank Stefan Müller-Stach, Morihiko Saito, Charles Vial and the referee for their comments on preliminary versions of the manuscript.

History of the origin of the book and acknowledgments (by the first author):

This book stems from lectures on pure motives that I have given over a period of years, starting with a lecture in 1988 at the University of Tokyo and followed by lectures – or series of lectures – at many other places, of which I mention in particular the lectures at the 2001 summer school in Grenoble (see the proceedings

[**MüS-Pe**]) and those at the Tata Institute of Fundamental Research in Mumbai in 2008.

I would like to thank the numerous people who have helped me before, during and after these lectures with advice and comments. Especially important and valuable for me have been the many discussions on algebraic cycles and motives over the years with Spencer Bloch and Uwe Jannsen and later, during our joint work [**Kahn-M-P**], with Bruno Kahn and Claudio Pedrini.

Finally, last but certainly not least, I would like to thank my two coauthors of the book. Without their help and encouragement these lectures would never have been transformed into a book; their help has been indispensable!

CHAPTER 1

Algebraic Cycles and Equivalence Relations

In this chapter k is an arbitrary field (possibly not algebraically closed) and $\mathsf{SmProj}(k)$ denotes the category of smooth projective varieties over k. A variety is a reduced scheme, not necessarily irreducible; indeed it is crucial to allow (finite) disjoint unions of irreducible ones. If we write X_d, we are dealing with an irreducible variety of dimension d.

Remark. The results in this Chapter are of course well known. They are collected at this place for the reader's convenience and, at the same time, to give us the opportunity to introduce some notation we use in the remainder of this book.

We would like to mention the following sources where the reader can find further details. For a general introduction to algebraic cycles, Chow groups and intersection theory: see [**Harts**, Appendix A], [**Voi03**, Part III]. For a more elaborate study we refer to [**Ful**], especially [**Ful**, Chapters 1, 6, 7, 8, 16]. An introduction to étale cohomology can be found in [**Harts**, Appendix] or [**Mil98**], while [**Mil80**] provides more details.

1.1. Algebraic Cycles

An *algebraic cycle* on a variety X is a formal finite integral linear combination $Z = \sum n_\alpha Z_\alpha$ of irreducible subvarieties Z_α of X. If all the Z_α have the same codimension i we say that Z is a codimension i cycle. We introduce the abelian group

$$\boxed{\mathsf{Z}^i(X) = \{\text{codim } i \text{ cycles on } X\}.}$$

If we prefer to think in terms of *dimension* and $X = X_d$ is pure dimensional, we write $\mathsf{Z}_{d-i}(X) = \mathsf{Z}^i(X)$. If, instead of integral linear combinations we work with coefficients in a field F (mostly, we use $F = \mathbb{Q}$), we write

$$\mathsf{Z}^i(X)_F := \left\{\sum r_\alpha Z_\alpha \mid r_\alpha \in F, \ Z_\alpha \text{ a codim. } i \text{ cycle}\right\} = \mathsf{Z}^i(X) \otimes_{\mathbb{Z}} F.$$

EXAMPLES 1.1.1.　　(1) The group of codimension 1 cycles, the *divisors*, is also written $\mathsf{Div}(X)$.

(2) The zero-cycles $\mathsf{Z}^d(X) = \mathsf{Z}_0(X)$. These are finite formal sums $Z = \sum_\alpha n_\alpha P_\alpha$ where P_α is an irreducible 0-dimensional k-variety. The *degree* of P_α is just the degree of the field extension $[k(P_\alpha)/k]$ and $\deg(Z) = \sum n_\alpha \deg(P_\alpha)$.

(3) To any subscheme Y of X with irreducible components Y_α of dimension d_α one can associate the class $[Y] = \sum_\alpha n_\alpha Y_\alpha \in \sum_\alpha \mathsf{Z}_{d_\alpha}(X)$, where n_α is the length of the zero-dimensional Artinian ring $\mathcal{O}_{Y,Y_\alpha}$; see [**Ful**, 1.5].

1

We recall the following operations on cycles. There are three basic operations: cartesian product, pushforward and intersection. The remaining ones can be deduced from these basic operations.

Cartesian product: The usual cartesian product of subvarieties can be linearly extended to products of cycles.

Push forward: If $f : X \to Y$ is a morphism of k-varieties and $Z \subset X$ is an irreducible subvariety, one sets

$$\deg(Z/f(Z)) = \begin{cases} [k(Z) : k((f(Z)))] & \text{if } \dim f(Z) = \dim Z \\ 0 & \text{if } \dim f(Z) < \dim Z \end{cases}$$

Intuitively, the number $[k(Z) : k((f(Z)))]$ gives the number of sheets of Z over $f(Z)$. Setting

$$f_*(Z) = \deg(Z/f(Z))f(Z)$$

and extending linearly, one gets a homomorphism

$$f_* : \mathsf{Z}_k(X) \to \mathsf{Z}_k(Y).$$

Intersection: Two subvarieties V and W of X with codimension say i and j intersect each other in a union of subvarieties Z_α, of (various) codimension $\geq i + j$ if X is smooth. See e.g. [**Harts**, p. 48]. We say that the intersection is *proper* if equality holds for all $Z = Z_\alpha$. If this is the case, the *intersection number* is defined by

$$i(V \cdot W; Z) := \sum_r (-1)^r \ell_A(\mathrm{Tor}_r^A(A/\mathcal{I}(V), A/\mathcal{I}(W))), \quad A = \mathcal{O}_{X,Z}$$

where $\mathcal{I}(V)$ denotes the ideal of the variety V in the ring A. The intersection product incorporates these:

$$V \cdot W = \sum_\alpha i(V \cdot W; Z_\alpha)Z_\alpha. \tag{1}$$

See [**Serre57**, P. 144]. This coincides with the earlier definitions due to Chevalley [**Chev58**], Weil [**Weil46**] and Samuel [**Sam58**].

Pull back: Let $f : X \to Y$ be a morphism in $\mathsf{SmProj}(k)$ and let $Z \subset Y$ be any subvariety. The graph of f is a subvariety $\Gamma_f \subset X \times Y$ and if it meets $X \times Z$ properly, we set

$$f^*Z := [\mathrm{pr}_X]_*(\Gamma_f \cdot (X \times Z)),$$

where $\mathrm{pr}_X : X \times Y \to X$ is the projection.[1]

Using the notion of pull back, we obtain an equivalent definition of the intersection product which is due to Weil [**Weil46**]:

$$\left. \begin{array}{rl} V \cdot W & = \quad \Delta_X^*(V \times W), \quad V, W \in \mathsf{Z}(X) \\ \Delta_X & : \quad X \hookrightarrow X \times X \text{ (the diagonal embedding).} \end{array} \right\} \tag{2}$$

---

[1]If f is flat the cycle f^*Z is nothing but the inverse image scheme $f^{-1}Z$, which is always defined; this definition can be extended linearly to cycles and yields a homomorphism $f^* : \mathsf{Z}^i(Y) \to \mathsf{Z}^i(X)$. See [**Ful**, §1.7].

Correspondences: A *correspondence* from X to Y is simply a cycle in $X \times Y$. A correspondence $Z \in \mathsf{Z}^t(X \times Y)$ acts on cycles on X as follows:

$$Z(T) \quad = \quad [\mathrm{pr}_Y]_*(Z \cdot (T \times Y)) \in \mathsf{Z}^{i+t-d}(Y), \tag{3}$$

$$T \in \mathsf{Z}^i(X), \ d = \dim(X),$$

whenever this is defined. If $t = d$ the correspondence preserves the codimension of the cycle and for this reason one calls $t - d$ the *degree* of the correspondence.

The *transpose* $^{\mathsf{T}}Z$ of Z is the same cycle, but considered as a subvariety of $Y \times X$.[2] The pull back operation f^* is the special case for $Z = {}^{\mathsf{T}}\Gamma_f$ and the operation f_* is the special case $Z = \Gamma_f$. Note that f^* has degree 0 while f_* has degree $\dim(Y) - \dim(X)$.

The operations of intersections, pull back and action of correspondences are not always defined. Therefore we search for a "good" equivalence relation on cycles which ensures that these operations are defined on the equivalence classes. This is the subject of the next section (§ 1.2).

1.2. Equivalence Relations

We shall use the shorthand "equivalence relation" for a family of equivalence relations on the groups $\mathsf{Z}(X) := \bigoplus_i \mathsf{Z}^i(X)$ i.e., for every variety X we have on every group $\mathsf{Z}^i(X)$ an equivalence relation. Such an equivalence relation $\sim$ is called *adequate* (or "good") if restricted to the category $\mathsf{SmProj}(k)$ it has the following properties [**Sam58**]:

R1: compatibility with grading and addition;
R2: compatibility with products: if $Z \sim 0$ on X, then for all Y one has $Z \times Y \sim 0$ in $\mathsf{Z}(X \times Y)$;
R3: compatibility with intersection: if $Z_1 \sim 0$ and $Z_1 \cdot Z_2$ is defined, then $Z_1 \cdot Z_2 \sim 0$;
R4: compatibility with projections: if $Z \sim 0$ on $X \times Y$, then $(\mathrm{pr}_X)_*(Z) \sim 0$ on X;
R5: *moving lemma*: given $Z, W_1, \ldots, W_\ell \in \mathsf{Z}(X)$ there exists $Z' \sim Z$ such that $Z' \cdot W_j$ is defined for $j = 1, \ldots, \ell$.

Having such a family of equivalence relations on the groups $\mathsf{Z}^i(X)$, we put

$$\mathsf{Z}^i_\sim(X) \quad = \quad \left\{ Z \in \mathsf{Z}^i(X) \mid Z \sim 0 \right\}.$$

This is a subgroup of $\mathsf{Z}^i(X)$ as follows from R1. If we now work with coefficients in a field F, we take

$$\mathsf{Z}^i_\sim(X)_F = \mathsf{Z}^i_\sim(X) \otimes_{\mathbb{Z}} F.$$

This is a vector subspace of $\mathsf{Z}^i(X_F)$. Now put

$$C^i_\sim(X) \quad := \quad \mathsf{Z}^i(X)/\mathsf{Z}^i_\sim(X), \quad C_\sim(X) = \bigoplus_i C^i_\sim(X),$$

$$C^i_\sim(X)_F \quad = \quad C^i_\sim(X) \otimes_{\mathbb{Z}} F, \quad C_\sim(X)_F = C_\sim(X) \otimes_{\mathbb{Z}} F.$$

The axioms are set up such that the following result is a consequence [**Sam58**, Prop. 6 and 7] :

---

[2]More precisely, $^{\mathsf{T}}Z = \tau_* Z$ where $\tau : X \times Y \to Y \times X$ interchanges the factors. So, even if $X = Y$ these two cycles are different in general.

LEMMA 1.2.1. *For any adequate equivalence relation $\sim$ on $X \in \mathsf{SmProj}(k)$ we have*

(1) *$C_\sim(X)$ is a ring with product induced from intersection of cycles;*
(2) *For any morphism $f : X \to Y$ in $\mathsf{SmProj}(k)$ the maps f_* and f^* induce (well defined) homomorphisms $f_* : C_\sim(X) \to C_\sim(Y)$ and $f^* : C_\sim(X) \to C_\sim(Y)$. The latter is a homomorphism of graded rings;*
(3) *a correspondence Z from X to Y of degree r induces $Z_* : C_\sim^i(X) \to C_\sim^{i+r}(Y)$ and equivalent correspondences induce the same Z_*.*

For instance, the fact that f^* respects intersections follows from formula (2):

$$f^*(V \cdot W) = f^* \Delta_Y^*(V \times W) = \Delta_X^*(f^*V \times f^*W) = f^*V \cdot f^*W, \quad V, W \in \mathsf{Z}(X).$$

We now discuss the most important equivalence relations.

A. Rational Equivalence. This is a generalization of the well known notion of linear equivalence for divisors. For the earliest, by now classical, results we refer the reader to [**Chow56, Sam56, Chev58**].

Rational equivalence on *any* variety X can be defined as follows [**Ful**, Chapter 1]. To start, the divisor of a rational function $f \in k(X)$ is defined as follows:

$$\operatorname{div}(f) := \sum_{Y \subset X} \operatorname{ord}_Y(f) \cdot Y, \quad Y \text{ of codim } 1,$$

where the (order) homomorphism $\operatorname{ord}_Y : k(X)^* \to \mathbb{Z}$ is defined as follows. The local ring $A = \mathcal{O}_{X,Y}$ is one dimensional and for $f \in A$ one puts $\operatorname{ord}_Y(f) = \ell_A(A/(f))$, with ℓ_A the *length* of an A-module. For $f \in k(X)^*$, write $f = \dfrac{f_1}{f_2}$ and put $\operatorname{ord}_Y(f) = \operatorname{ord}_Y(f_1) - \operatorname{ord}_Y(f_2)$. This is well-defined, which is easy to see, e.g. [**Ful**, Appendix A].

It follows that the divisor $\operatorname{div}(f)$ for a function $f \in K(Y)^*$ on an irreducible subvariety $Y \subset X$ is a codimension 1 cycle on Y and hence, if Y is of codimension $i - 1$ in X, $\operatorname{div}(f) \in \mathsf{Z}^i(X)$ and by definition $\mathsf{Z}_{\mathrm{rat}}^i(X)$ is the subgroup generated by such cycles. In other words, for a codimension i-cycle one has $Z \sim_{\mathrm{rat}} 0$ if and only if there is a finite collection of pairs (Y_α, f_α) of codimension $(i-1)$ irreducible varieties and non-zero functions on them such that $Z = \sum \operatorname{div}(f_\alpha)$. Equivalently, if $X^{(i)}$ stands for the collection of irreducible codimension i subvarieties of X we have

$$\mathsf{Z}_{\mathrm{rat}}^i(X) = \operatorname{Im}\left\{ \bigoplus_{Y \in X^{(i-1)}} k(Y)^* \xrightarrow{\operatorname{div}} \bigoplus_{Z \in X^{(i)}} \mathbb{Z} \right\}$$

The *Chow groups* are the cokernels of these maps:

$$\begin{aligned} \mathsf{CH}^i(X) &:= \mathsf{Z}^i(X)/\mathsf{Z}_{\mathrm{rat}}^i(X) = C_{\mathrm{rat}}^i(X) \\ \mathsf{CH}(X) &:= \bigoplus_i \mathsf{CH}^i(X). \end{aligned}$$

EXAMPLE 1.2.2. For divisors (codimension one cycles), rational equivalence indeed coincides with linear equivalence. The quotient of the group $\mathsf{Div}(X)$ by the subgroup $\{\operatorname{div}(f) \mid f \in k(X)^\times\}$ (divisors *linearly equivalent to* 0) is the *Picard group* $\mathsf{Pic}(X)$ if X is smooth.

If we work with coefficients in a field F we shall write

$$\mathsf{CH}^i(X)_F = C_{\mathrm{rat}}^i(X)_F = C_{\mathrm{rat}}^i(X) \otimes_{\mathbb{Z}} F, \quad \mathsf{CH}(X)_F = \bigoplus_i \mathsf{CH}^i(X)_F$$

respectively. We stress at this point that the Chow groups $\mathsf{CH}^i(X)$ are defined for *any* variety X. But for smooth projective varieties we know more is true:

PROPOSITION 1.2.3. *On the category* $\mathsf{SmProj}(k)$ *rational equivalence is an adequate equivalence relation. So, if X is smooth projective the direct sum* $\mathsf{CH}(X)$ *is in fact a graded ring (and* $\mathsf{CH}(X)_F$ *a graded F-algebra).*

Let us make a few remarks on the proof. R1 is obvious and R2 and R3 are easy. The proof of R4 follows from the more general

PROPOSITION 1.2.4. *Let $f : X \to Y$ be a proper morphism between varieties (not necessarily smooth) and let Z be a cycle rationally equivalent to zero. Then* $f_* Z \sim_{\mathrm{rat}} 0$.

For a proof see [**Ful**, p. 12]. The idea here is as follows. If $Z = \mathrm{div}(\varphi)$ where φ is a function on a subvariety of X, we may replace X by this subvariety and Y by $f(X)$ so that f is proper and surjective and Z a divisor of a function on X. If f is equidimensional, one uses the *norm* (=the determinant of the linear map given by multiplication with φ)

$$N : k(X) \to k(Y).$$

Indeed, one has $f_* Z = f_*(\mathrm{div}(\varphi)) = \mathrm{div}(N(\varphi))$. If f is not equidimensional, $f_* Z = 0$. The general case can be reduced to this situation.

This method of proof can be used to give an *alternative definition of rational equivalence*[3] as follows:

LEMMA 1.2.5. *Suppose X is smooth and projective. Then $Z \in \mathsf{Z}^i(X)$ is rationally equivalent to zero if and only if there exists a cycle $W \in \mathsf{Z}^i(X \times \mathbb{P}^1)$ and $a, b \in \mathbb{P}^1$ such that, defining*

$$W(t) := (\mathrm{pr}_X)_* \left(W \cdot (X \times t) \right),$$

we have $W(a) = 0$ and $W(b) = Z$.

Finally, about the moving lemma (R5). We sketch the proof of [**Rob**]. The situation is as follows. We have $X \in \mathsf{SmProj}(k)$ such that $X \hookrightarrow \mathbb{P}^N$, some fixed projective space. We may assume that Z is a codimension i subvariety of X, that $\ell = 1$ and that $W = W_1$ is a subvariety of X of codimension j. If $\mathrm{codim}(W \cap Z) = i + j$ the intersection is proper and the Lemma is trivially true by taking $Z' = Z$. If not, put $\mathrm{excess}(W \cap Z) = \mathrm{codim}(W \cap Z) - (i + j) > 0$. The proof consists of two steps:

(1) If $X = \mathbb{P}^N$ one shows that a generic linear transformation moves Z to a cycle which intersects W properly.

(2) In the general case, take a "general" linear space $L \subset \mathbb{P}^N$ of codimension $d+1$. Consider the cone $C(L, Z)$ on Z with L as vertex. Since L is general, the intersection $C(L, Z) \cdot X$ is well defined and in fact, by construction, one has

$$C(L, Z) \cdot X = Z + Z_*.$$

For general enough L one shows that $\mathrm{excess}(Z_* \cap W) < \mathrm{excess}(Z \cap W)$. By step (1), for a general linear transformation τ the intersection $\tau(C(L, Z) \cdot X) \cdot W$ is defined and since

$$Z \sim_{\mathrm{rat}} \tau(C(L, Z) \cdot X) - Z_*,$$

---

[3]This is the definition used by Chow and Samuel.

one can proceed by induction on the excess.

We end this discussion on rational equivalence by stating the main properties of Chow groups.

Theorem 1.2.6.   (1) *If $X \in \mathsf{SmProj}(k)$, then the Chow group $\mathsf{CH}(X) = \bigoplus \mathsf{CH}^q(X)$ is a graded commutative ring with respect to intersection product;*

(2) *if $f : X \to Y$ be a morphism in $\mathsf{SmProj}(k)$, then $f^* : \mathsf{CH}(Y) \to \mathsf{CH}(X)$ is a graded ring homomorphism, while $f_* : \bigoplus \mathsf{CH}_q(X) \to \bigoplus \mathsf{CH}_q(Y)$ is an additive graded homorphism of degree $\dim Y - \dim X$.*

(3) *If $X, Y \in \mathsf{SmProj}(k)$, then $Z \in \mathsf{Corr}^e_{\mathrm{rat}}(X, Y)$ induces an additive homomorphism $Z_* : \mathsf{CH}(X) \to \mathsf{CH}(Y)$ of degree e.*

Suppose that, more generally, X, Y are arbitrary k–varieties. Then

(4) *If $i : Y \hookrightarrow X$ is a closed embedding and $j : U = X - Y \hookrightarrow X$, one has an exact sequence*

$$\mathsf{CH}_q(Y) \xrightarrow{i_*} \mathsf{CH}_q(X) \xrightarrow{j^*} \mathsf{CH}_q(U) \to 0.$$

(5) *The homotopy property holds: the projection $\mathrm{pr}_X : X \times \mathbb{A}^n \to X$ induces an isomorphism $\mathrm{pr}_X^* : \mathsf{CH}^i(X) \xrightarrow{\sim} \mathsf{CH}^i(X \times \mathbb{A}^n)$.*

Proof: Properties (1) and (2) follow immediately from the fact that $\sim_{\mathrm{rat}}$ is a good equivalence relation (cf. Prop. 1.2.3) and (3) then follows from the preceding formula (3). Property (4) is straightforward and for (5) see [**Ful**, Page 22].   □

Remark. In [**Ful**, Chapter 6] it is shown that one can define a refined intersection product on $\mathsf{CH}^*(X)$ without using the moving lemma (by deformation to the normal cone). If X is smooth, this product satisfies all the good properties mentioned above (cf. Theorem 1.2.6). The refined intersection product has the advantage that we can keep control over the support of the intersection class; we shall use this fact (without saying so) in Chapter 7. Moreover, in certain cases it can be used to intersect cycles on singular varieties; this will be used in Chapter 8.

B. Algebraic Equivalence. Now we suppose that X is smooth projective. For the definition of algebraic equivalence, we go back to the alternative definition for rational equivalence as given in Lemma 1.2.5 and we replace $\mathbb{P}^1$ by any curve:

Definition 1.2.7 ([**Weil54**]). $Z \sim_{\mathrm{alg}} 0$ if there is a smooth irreducible curve C and $W \in \mathsf{Z}^i(C \times X)$ and two points $a, b \in C$ such that $W(a) = 0$, $W(b) = Z$. One puts

$$
\begin{aligned}
\mathsf{Z}^i_{\mathrm{alg}}(X) \quad &:= \quad \left\{ Z \in \mathsf{Z}^i(X) \mid Z \sim_{\mathrm{alg}} 0 \right\} \\
\mathsf{CH}^i_{\mathrm{alg}}(X) \quad &:= \quad \frac{\mathsf{Z}^i_{\mathrm{alg}}(X)}{\mathsf{Z}^i_{\mathrm{rat}}(X)} \subset \mathsf{CH}^i(X).
\end{aligned}
$$

Remarks 1.2.8.   (1) Instead of C we could have taken any smooth variety V since, given $a, b \in V$ there is always a smooth curve $C \subset V$ passing through a and b: take a sufficiently general linear section containing a and b.

(2) We could also replace C by an abelian variety as in [**Weil54**, lemma 9 p.108].

(3) One clearly has $\mathsf{Z}^i_{\mathrm{rat}}(X) \subset \mathsf{Z}^i_{\mathrm{alg}}(X)$ but in general the two are distinct: take for X an elliptic curve and a, b two distinct points. Then $Z = a - b$ is not the divisor of a function, but Z is algebraically equivalent to zero.

(4) The Chow group $\mathsf{CH}^i(X)$ surjects onto

$$\frac{\mathsf{Z}^i(X)}{\mathsf{Z}^i_{\mathrm{alg}}(X)} = \frac{\mathsf{CH}^i(X)}{\mathsf{CH}^i_{\mathrm{alg}}(X)}$$

which is a countable group by the theory of Chow varieties [**Chow-vW**].

C. Smash Nilpotent Equivalence. Again, X is supposed to be smooth projective. First some more notation. For a variety X and a cycle Z on X we set

$$X^n \; := \; \underbrace{X \times \cdots \times X}_{n}$$

$$Z^n \; := \; \underbrace{Z \times \cdots \times Z}_{n}.$$

DEFINITION 1.2.9 ([**Voe00**]). $Z \sim_\otimes 0$ if and only if for some positive integer n one has $Z^n \sim_{\mathrm{rat}} 0$ on X^n.

PROPOSITION 1.2.10. *Smash-nilpotent equivalence is a good equivalence relation. In particular,* $\mathsf{Z}^i_\otimes(X) := \big\{ Z \in \mathsf{Z}^i(X) \mid Z \sim_\otimes 0 \subset \mathsf{Z}^i(X) \big\}$ *is a subgroup of* $\mathsf{Z}^i(X)$.

Proof: The required properties are straightforward to check. As an example, let us verify that the set $\mathsf{Z}^i_\otimes(X) \subset \mathsf{Z}^i(X)$ is a subgroup. Take two cycles $Z, Z' \in \mathsf{Z}^i_\otimes(X)$ and consider

$$(Z + Z')^n = \sum_{r+s=n} \binom{n}{r} Z^r \times Z'^s \tag{4}$$

inside $\mathsf{CH}^{in}(\underbrace{X \times \cdots \times X}_{n})$ (after a suitable reordering of the factors). If $n \gg 0$ we have that either the first or the second factor is rational equivalent to zero so that also $Z^r \times Z'^s \sim_{\mathrm{rat}} 0$ and hence $Z + Z' \sim_\otimes 0$. $\qquad\square$

COROLLARY 1.2.11. *Smash-nilpotence is preserved by correspondences.*

An important result, due independently to Voevodsky and Voisin, is the following.

THEOREM 1.2.12 ([**Voe95, Voi96**]). [4] *One has* $\mathsf{Z}^i_{\mathrm{alg}}(X)_\mathbb{Q} \subset \mathsf{Z}^i_\otimes(X)_\mathbb{Q}$.

We shall give a sketch of the proof in Appendix B.

D. Homological Equivalence. Let F be a field of characteristic 0 and let GrVect_F be the category of finite dimensional graded F-vector spaces.

DEFINITION 1.2.13. A *Weil-cohomology theory* is a functor

$$H : \mathsf{SmProj}(k)^{\mathrm{opp}} \to \mathrm{GrVect}_F$$

which satisfies the following axioms[5]

   (1) there exists a cup product $\cup : H(X) \times H(X) \to H(X)$ which is graded and super-commutative , i.e. if $a \in H^i(X)$, $b \in H^j(X)$, then $b \cup a = (-1)^{ij} a \cup b$.;

---

[4] In [**Voi96**] the result is stated only for cycles on self-products $X \times X$

[5] This is the convention of [**Klei94**]. There are definitions that are less restrictive, see e.g. [**Klei68**]. However, the classical Weil cohomology theories satisfy the properties given here. For simplicity we omit Tate twists; see [**DJ**] for the precise version.

(2) one has Poincaré duality:

there is a trace isomorphism $\mathrm{Tr} : H^{2d}(X_d) \xrightarrow{\sim} F$ such that

$$H^i(X_d) \times H^{2d-i}(X_d) \xrightarrow{\cup} H^{2d}(X_d) \xrightarrow{\mathrm{Tr}} F$$

is a perfect pairing (in particular, $H^0(\text{point}) \simeq F$);

(3) the Künneth formula holds:

$$H(X) \otimes H(Y) \xrightarrow{(\mathrm{pr}_X)^* \otimes (\mathrm{pr}_Y)^*} H(X \times Y) \tag{5}$$

is a graded isomorphism;

(4) there are *cycle class* maps

$$\gamma_X : \mathsf{CH}^i(X) \to H^{2i}(X) \tag{6}$$

which are

- functorial in the sense that for $f : X \to Y$ in $\mathsf{SmProj}(k)$, one has $f^* \circ \gamma_Y = \gamma_X \circ f^*$ and $f_* \circ \gamma_X = \gamma_Y \circ f_*$;
- compatible with intersection product

$$\gamma_X(\alpha \cdot \beta) = \gamma_X(\alpha) \cup \gamma_X(\beta); \tag{7}$$

- compatible with points P: this means that

$$\mathrm{Tr} \circ \gamma_P = \deg, \tag{8}$$

i.e., the following diagram commutes:

$$
\begin{array}{ccc}
\mathsf{CH}^0(P) & \xrightarrow{\gamma_P} & H^0(P) \\
{\scriptstyle \deg}\downarrow & & \downarrow{\scriptstyle \mathrm{Tr}} \\
\mathbb{Z} & \hookrightarrow & F.
\end{array}
$$

As a matter of notation, write

$$\boxed{\mathsf{A}^i(X) := \mathrm{Im}(\gamma_X) = H^{2i}_{\mathrm{alg}}(X).} \tag{9}$$

(5) Weak Lefschetz holds: if $i : Y_{d-1} \hookrightarrow X_d$ is a smooth hyperplane section, then

$$H^i(X) \xrightarrow{i^*} H^i(Y) \text{ is } \begin{cases} \text{an isomorphism} & \text{for } i < d-1 \\ \text{injective} & \text{for } i = d-1. \end{cases}$$

(6) Hard Lefschetz holds: the Lefschetz-operator $L(\alpha) = \alpha \cup \gamma_X(Y)$ induces isomorphisms

$$L^{d-i} : H^{d-i}(X) \xrightarrow{\sim} H^{d+i}(X), \quad 0 \le i \le d.$$

EXAMPLES 1.2.14 (Classical Weil cohomology theories).

(1) If $\mathrm{char}(k) = 0$ and $k \subset \mathbb{C}$ one can take

    (a) $H^i_{\mathrm{B}}(X)$, the *Betti-cohomology*, i.e. singular cohomology group $H^i(X_{\mathrm{an}})$ with $\mathbb{Q}$-coefficients $F = \mathbb{Q}$ or $\mathbb{C}$-coefficients ($F = \mathbb{C}$). Here X_{an} is the complex manifold underlying X;

    (b) the *classical De Rham cohomology* $H_{\mathrm{dR}}(X_{\mathrm{an}}; \mathbb{C})$ (here $F = \mathbb{C}$);

    (c) the *algebraic de Rham cohomology* $H^i_{\mathrm{dR}}(X) := \mathbb{H}^i(X_{\mathrm{Zar}}, \Omega^\bullet_{X/k})$ (with $F = k$).

That (a) gives a Weil cohomology is indeed classical: ordinary singular cohomology satisfies all of the above requirements; the classical de Rham theorem (for (b)), Grothendieck's version in the algebraic setting (for (c)), and the comparison theorems stated below (Remark 1.2.15) then imply that (b) and (c) also satisfy the requirements.

(2) For a variety X defined over a field k one can take the *étale cohomology* groups $H^i_{\text{ét}}(X_{\bar{k}}, \mathbb{Q}_\ell)$. Here $X_{\bar{k}} = X \otimes \bar{k}$, the variety considered over its algebraic closure and ℓ is any prime number different from the characteristic of k. Note that we follow the common *abuse of notation* for $H^i_{\text{ét}}(X_{\bar{k}}, \mathbb{Q}_\ell)$ (to be used throughout the book). For the reader's convenience we shall recall the definitions:

$$H^i_{\text{ét}}(X_{\bar{k}}, \mathbb{Z}_\ell) \quad := \quad \varprojlim_n H^i_{\text{ét}}(X_{\bar{k}}, \mathbb{Z}/\ell^n\mathbb{Z})$$

$$H^i_{\text{ét}}(X_{\bar{k}}, \mathbb{Q}_\ell) \quad := \quad H^i_{\text{ét}}(X_{\bar{k}}, \mathbb{Z}_\ell) \otimes_{\mathbb{Z}_\ell} \mathbb{Q}_\ell.$$

Note that these groups (or to be precise: $\mathbb{Z}_\ell$–modules, resp. $\mathbb{Q}_\ell$–vector spaces) should not be confused with the cohomology groups of $X_{\bar{k}}$ with coefficients in the constant sheaf $\mathbb{Z}_\ell$ (resp. $\mathbb{Q}_\ell$) because for étale cohomology one must use torsion sheaves.

That $H^*_{\text{ét}}(X_{\bar{k}}, \mathbb{Q}_\ell)$ indeed gives a Weil cohomology is not trivial at all; in particular the hard Lefschetz property is very difficult (see [**Del80**]). Further details on étale cohomology can be found in [**Mil80**].

(3) For k perfect, one has *crystalline cohomology* $H^i_{\text{crys}}(X/W(k)) \otimes K$ where K is the field of fractions of the Witt ring $W(k)$. Except for the hard Lefschetz property, the proof that this is a Weil cohomology is a consequence of the work of Berthelot [**Ber**]; the hard Lefschetz property has been deduced by Katz and Messing [**Katz-Me**] from the validity of the Weil conjectures (proven by Deligne [**Del74b, Del80**]). See the excellent overview [**Ill**].

Remark 1.2.15. Some of these cohomology theories are compatible in the sense that there are well-defined canonical isomorphisms as follows.
1) If $k \subset \mathbb{C}$ one has the *de Rham theorem*

$$H_{dR}(X) \otimes_k \mathbb{C} \xrightarrow{\sim} H_{dR}(X_{\text{an}}; \mathbb{C}) \xrightarrow{\sim} H_{\text{B}}(X) \otimes_{\mathbb{Z}} \mathbb{C}.$$

See [**Zar**, Mumford, Appendix to Chap. VII].
2) If $k = \mathbb{C}$ one has the *theorem of Artin* [**SGA4**, XI, thm. 4.4., p. 75]:

$$H_{\text{ét}}(X, \mathbb{Q}_\ell) \xrightarrow{\sim} H_{\text{B}}(X) \otimes_{\mathbb{Q}} \mathbb{Q}_\ell.$$

Fix now a Weil cohomology theory and use the cycle class map (6) to make the following

DEFINITION 1.2.16.

$$Z \sim_{\text{hom}} 0 \iff \gamma_X(Z) = 0.$$

(Note that this depends on the choice of a Weil cohomology theory[6].)

---

[6]But if Conjecture D (1.2.19) would hold, then homological equivalence would be independent of the choice

E. Numerical equivalence.

DEFINITION 1.2.17. Let $X_d \in \mathsf{SmProj}(k)$. For $Z \in \mathsf{Z}^i(X_d)$ we put $Z \sim_{\mathrm{num}} 0$ if for every $W \in \mathsf{Z}^{d-i}(X_d)$ such that $Z \cdot W$ is defined (and hence a zero-cycle) we have $\deg(Z \cdot W) = 0$.

The following lemma describes how numerical equivalence relates to other adequate relations.

LEMMA 1.2.18. *We have*
  (i) $\mathsf{Z}^i_{\mathrm{alg}}(X) \subset \mathsf{Z}^i_{\mathrm{hom}}(X)$
  (ii) $\mathsf{Z}^i_{\otimes}(X) \subset \mathsf{Z}^i_{\mathrm{hom}}(X)$
  (iii) $\mathsf{Z}^i_{\mathrm{hom}}(X) \subset \mathsf{Z}^i_{\mathrm{num}}(X)$.

Proof: Part (i) follows from the observation

$$(a) - (b) \sim_{\mathrm{hom}} 0 \implies Z(a) - Z(b) = \mathrm{pr}_X(Z \cdot ((a) - (b))) \sim_{\mathrm{hom}} 0.$$

For part (ii), note that if $Z^n \sim 0$ then its cycle class

$$\gamma_{X^n}(Z^n) = \underbrace{\gamma_X(Z) \otimes \cdots \otimes \gamma_X(Z)}_{n}$$

is zero in $H^{2in}(X^n)$, and hence $\gamma_X(Z) = 0$. Part (iii) follows from the compatibility of the cycle class map with intersection products and points: for a zero-cycle Z we have $\deg(Z) = \mathrm{Tr}\,\gamma_X(Z)$ by property (8) of a Weil cohomology theory, hence homological and numerical equivalence coincide for zero-cycles. If $Z \in \mathsf{Z}^i_{\mathrm{hom}}(X)$ with $i < d$ and $W \in \mathsf{Z}^{d-i}(X)$ then

$$\deg(Z \cdot W) = \mathrm{Tr}(\gamma_X(Z) \cup \gamma_X(W)) = 0$$

by property (6). $\square$

We have the fundamental

CONJECTURE 1.2.19 (Conjecture $D(X)$). Suppose $k = \bar{k}$. Then

$$\mathsf{Z}^i_{\mathrm{hom}}(X) = \mathsf{Z}^i_{\mathrm{num}}(X).$$

This conjecture is known for divisors ($i = 1$) in arbitrary characteristic (theorem of Matsusaka, see Appendix A); furthermore, in characteristic zero it is also known for $i = 2$, for dimension 1, and for abelian varieties [**Liebm**].

Furthermore, one has:

THEOREM 1.2.20 ([**Klei68**, thm. 3.5]). *Suppose that $k = \bar{k}$ (any characteristic). Then*

$$\mathrm{Nm}^i(X)_{\mathbb{Q}} := C^i_{\mathrm{num}}(X)_{\mathbb{Q}} = [\mathsf{Z}^i(X)/\mathsf{Z}^i_{\mathrm{num}}(X)] \otimes_{\mathbb{Z}} \mathbb{Q}$$

is a finite dimensional $\mathbb{Q}$–vector space of dimension $\leq b_{2i}(X) = \dim_{\mathbb{Q}_\ell} H^{2i}_{\text{ét}}(X)$.

In fact, we shall show this in chapter 3 (Lemma 3.2.2).

Another important conjecture is:

CONJECTURE 1.2.21 (Voevodsky's Conjecture [**Voe00**]). Suppose that $k = \bar{k}$ Then $\mathsf{Z}^i_{\otimes}(X) = \mathsf{Z}^i_{\mathrm{num}}(X)$.

Note that this conjecture is *independent* of the choice of a Weil cohomology theory, but it implies the standard conjecture $D(X)$ for *every* Weil cohomology theory since, by Lemma 1.2.18 (iii), we have

$$\mathsf{Z}^i_{\otimes}(X) \subset \mathsf{Z}^i_{\mathrm{hom}}(X) \subset \mathsf{Z}^i_{\mathrm{num}}(X) \subset \mathsf{Z}^i(X).$$

1.2.1. Summary: the Different Equivalence Relations for Smooth Projective Varieties. It can be shown that rational equivalence is the finest good equivalence relation [**Sam58**, Prop. 8], and numerical equivalence is the coarsest one.

We have a chain of inclusions

$$\boxed{Z^i_{\mathrm{rat}}(X) \subset Z^i_{\mathrm{alg}}(X) \subset Z^i_{\mathrm{hom}}(X) \subseteq Z^i_{\mathrm{num}}(X) \subsetneq Z^i(X).}$$

The first inclusion can be strict (see Remark 1.2.8). If $i > 1$, we may have $Z^i_{\mathrm{alg}}(X) \neq Z^i_{\mathrm{hom}}(X)$ as was shown by Griffiths [**Griff69**]; see also Appendix A, subsection A-3.2. For $\mathbb{Q}$–coefficients we have:

$$Z^i_{\mathrm{alg}}(X)_{\mathbb{Q}} \subset Z^i_{\otimes}(X)_{\mathbb{Q}} \subseteq Z^i_{\mathrm{hom}}(X)_{\mathbb{Q}} \subseteq Z^i_{\mathrm{num}}(X)_{\mathbb{Q}}$$

If $k = \bar{k}$ the last two inclusions are expected to be equalities. In any case, the following chain of inclusions of subgroups of the Chow group is a consequence:

$$\boxed{\mathsf{CH}^i_{\mathrm{alg}}(X) \subset \mathsf{CH}^i_{\mathrm{hom}}(X) \subseteq \mathsf{CH}^i_{\mathrm{num}}(X) \subsetneq \mathsf{CH}^i(X).}$$

Appendix A: Survey of Some of the Main Results on Chow Groups

In this Appendix k is an algebraically closed field and and $X \in \mathsf{SmProj}(k)$.

From now on we shall use the word *correspondence* to mean *correspondence class* and use the notation (11), (12) from Section 2.1.

A-1. Divisors

We have

$$\boxed{\mathsf{CH}^1(X) = \mathsf{Div}(X)/\text{linear equivalence} = \mathrm{Pic}(X)}\,,$$

the Picard group. Moreover, this group is isomorphic to the group

$$H^1_{\mathrm{Zar}}(X, \mathcal{O}_X^\times) = H^1_{\text{ét}}(X, \mathcal{O}_X^\times)$$

(see [**Mil80**, Thm. 4.9]) of isomorphism classes of line bundles through the isomorphism induced by the homomorphism $D \mapsto \mathcal{O}_X(D)$.

THEOREM. *The group* $\mathsf{CH}^1_{\mathrm{alg}}(X)$ *is isomorphic to the group* $\mathrm{Pic}^0_{\mathrm{red}}(X)(k)$, *the reduced scheme associated to the component of the identity of the Picard scheme of* X. *This is an abelian variety.*

Classically, for $k = \mathbb{C}$ this goes back, in a less precise way, to Castelnuovo, Picard and Painlevé [**Zar**, p. 104]. In the algebraic setting for arbitrary fields k and in a much more precise fashion, this has been proved by Matsusaka [**Mats52**], Weil (in 1954; see [**La**, Chapter IV]), Chow [**Chow54**], Chevalley [**Chev58**], Seshadri [**Sesh62**]. It is now part of a much more general theory, the theory of the *Picard functor* due to Grothendieck, [**Groth62**]. See also [**Klei05**] and [**Bo-Lu-Ra**].

Concerning numerical equivalence we have:

THEOREM (Matsusaka's Theorem [**Mats57**]). $\mathsf{Div}_{\mathrm{hom}}(X) = \mathsf{Div}_{\mathrm{num}}(X)$. *In fact, these groups coincide with*

$$\boxed{\mathsf{Div}_\tau(X) := \left\{ D \in \mathsf{CH}^1(X) \mid nD \sim_{\mathrm{alg}} 0 \text{ for some } n \in \mathbb{Z} \right\}.}$$

And concerning algebraic equivalence:

THEOREM. *The* Néron-Severi group

$$\boxed{\mathsf{NS}(X) := \mathsf{Div}X/\mathsf{Div}_{\mathrm{alg}}(X)}$$

is a finitely generated abelian group.

Severi proved this for surfaces around 1908–09 for $k = \mathbb{C}$ (see e.g. [**Zar**, p. 107–112]) and Néron [**Ner**] in 1952 for any algebraically closed field.

Remark. In case $k = \mathbb{C}$ the above theorems are consequences of the long exact sequence in cohomology coming from the exponential sequence and GAGA [**Serre56**]. For details, see e.g. [**Griff-Ha, Mum74**].

A-2. Classical Results on the Picard and Albanese Varieties

For proofs consult [**La**]; see also [**Scholl**, p. 174]).

(a) For $X = X_d \in \mathsf{SmProj}(k)$ there exist two *abelian varieties*

$$\boxed{A_X = \mathrm{Alb}_X \quad \text{and} \quad P_X = (\mathrm{Pic}^0_X)_{\mathrm{red}}}$$

with the following *universal properties*:

(i) *Albanese variety* $A_X = \mathrm{Alb}_X$: fixing a point $x_0 \in X$, we have a morphism $\mathrm{alb}_X : X \to \mathrm{Alb}_X$ such that $\mathrm{alb}_X(x_0) = 0$ and such that alb_X is *universal* for morphisms to abelian varieties, i.e., given a morphism $f : X \to B$ to an abelian variety B with $f(x_0) = 0$, there exists a (unique) homomorphism of Abelian varieties $\overline{f} : \mathrm{Alb}_X \to B$ that makes the following diagram commute:

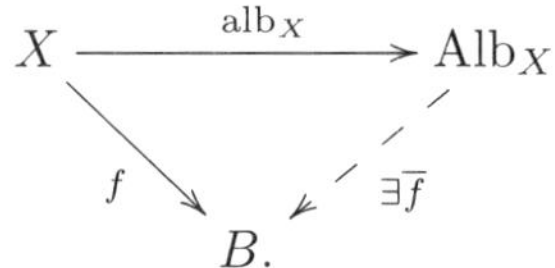

(ii) *Picard variety* $P_X = (\mathrm{Pic}^0_X)_{\mathrm{red}}$, where Pic_X is the Picard scheme of X, Pic^0_X is the connected component of the identity and the subscript "red" means that we take the underlying reduced scheme (which is – in our situation – an abelian variety). Fix moreover a point $x_0 \in X$. Then there exists a divisor class (the *Poincaré-divisor class*)

$$\mathcal{P}_X \in \mathsf{CH}^1(P_X \times X) = \mathsf{Corr}^{1-e}(P_X, X), \quad e = \dim P_X$$

normalized by[7]

$$\mathcal{P}_X(0) = 0, \quad {}^{\mathsf{T}}\mathcal{P}_X(x_0) = 0$$

such that for the pair $(P_X, \mathcal{P}_X)$ the following universal property holds: given a pair $(S, \mathcal{D})$ consisting of an irreducible variety $S \in \mathsf{SmProj}(k)$ and a divisor class $\mathcal{D} \in \mathsf{CH}^1(S \times X)$ normalized by ${}^{\mathsf{T}}\mathcal{D}(x_0) = 0$ and such that for some point $s_0 \in S$ the divisor $\mathcal{D}(s_0)$ belongs to $\mathsf{CH}^1_{\mathrm{alg}}(X)$ (and hence for all $s \in S$ the divisor $\mathcal{D}(s)$ belongs to $\mathsf{CH}^1_{\mathrm{alg}}(X)$), then there exists a unique morphism $\lambda : S \to P_X$ such that $\mathcal{D} = (\lambda \times \mathrm{id}_X)^*(\mathcal{P}_X)$. So we in fact have a bijection

$$\mathrm{Hom}(S, P_X) \xrightarrow{\sim} \left\{ (S, \mathcal{D}) \mid \mathcal{D} \in \mathsf{CH}^1(S \times X), {}^{\mathsf{T}}\mathcal{D}(x_0) = 0, \mathcal{D}(s) \in \mathsf{CH}^1_{\mathrm{alg}}(X) \right\}$$

that sends $\lambda : S \to P_X$ to the divisor class

$$\mathcal{D}(\lambda) = (\lambda \times \mathrm{id}_X)^*(\mathcal{P}_X).$$

Note that in particular, taking $S = \mathrm{Spec}\, k$, we get

$$P_X(k) = (\mathrm{Pic}^0_X)_{\mathrm{red}}(k) = \mathsf{CH}^1_{\mathrm{alg}}(X).$$

---

[7]Here one uses the action (Lemma 1.2.1) of the correspondence $\mathcal{P}_X$ and its transpose on zero-cycles which, in both cases produces divisors.

(b) *Divisorial correspondences.* See [**La**, p. 153]. Given X_d and Y_e in $\mathsf{SmProj}(k)$, fix points $e_X \in X$ and $e_Y \in Y$. From the above we get the isomorphisms

$$\mathrm{Hom}^0_{\mathsf{SmProj}(k)}(X, P_Y) \quad \xleftarrow{\cong} \quad \mathrm{Hom}_{\mathrm{AV}}(A_X, P_Y)$$

$$\xrightarrow[\Phi]{\cong} \quad A := \left\{ \begin{array}{c} D \in \mathsf{CH}^1(X \times Y) \\ D(e_X) = 0, \ {}^{\mathsf{T}}D(e_Y) = 0 \end{array} \right\}$$

where Hom^0 stands for pointed morphisms (i.e, morphisms that send the base point e_X to the origin 0) and Hom_{AV} for morphisms of abelian varieties.

By definition, the subgroup of *degenerate* divisors on $X \times Y$ is the subgroup generated by divisors D such that $\mathrm{pr}_X(D) \neq X$ or $\mathrm{pr}_Y(D) \neq Y$. A divisor D is degenerate if and only if $D = D_1 \times Y + X \times D_2$ with D_1 (resp. D_2) a divisor on X (resp. Y). Introduce the following notation:

$$\boxed{\mathsf{CH}^1_{\underline{\underline{\equiv}}}(X \times Y) : \text{the subgroup of } \mathsf{CH}^1(X \times Y) \text{ of classes of degenerate divisors.}}$$

Claim. The natural homomorphism

$$\Psi : A \to \mathsf{CH}^1(X \times Y)/\mathsf{CH}^1_{\underline{\underline{\equiv}}}(X \times Y)$$

induced by the inclusion $A \hookrightarrow \mathsf{CH}^1(X \times Y)$ is an isomorphism.

Proof: For the injectivity, note that if a divisor $D = D_1 \times Y + X \times D_2$ has the property $D(e_X) = 0$ and ${}^{\mathsf{T}}D(e_Y) = 0$ it must be the zero-divisor.

For the surjectivity, note that if necessary we can always add some degenerate divisor to a given divisor on $X \times Y$ so that for the resulting divisor D one has the required normalizations $D(e_X) = 0$ and ${}^{\mathsf{T}}D(e_Y) = 0$ (simply subtract $X \times D(e_X) + {}^{\mathsf{T}}D(e_Y) \times Y$ if necessary). $\qquad\square$

Denoting the jacobian of the curve C by $\mathrm{J}(C)$ as a special case we find back Weil's Theorem:

THEOREM A-2.1 (Weil's Theorem [**Weil48**, thm 22, Chap. VI]).

$$\mathrm{Hom}_{\mathrm{AV}}(\mathrm{J}(C), \mathrm{J}(C')) \xrightarrow[\Psi \circ \Phi]{\sim} \mathsf{CH}^1(C \times C')/\mathsf{CH}^1_{\underline{\underline{\equiv}}}(C \times C').$$

Remark. This result goes back to italian geometry [**Weil48**, p. 78].

(c) The above constructions are functorial [**Scholl**, Prop.3.10, p. 175]. More precisely, if $\lambda : X' \to X$ and $\mu : Y' \to Y$ are morphisms of varieties, then – with the notations of (b) above – if $\beta : A_X \to P_Y$ is a homomorphism of abelian varieties, we have $\Phi(\beta \circ \mathrm{alb}(\lambda)) = \Phi(\beta) \circ \Gamma_\lambda = \Phi(\beta) \circ \lambda_*$ and $\Phi(\mathrm{pic}(\mu) \circ \beta) = {}^{\mathsf{T}}\Gamma_\mu \circ \Phi(\beta) = \mu^* \circ \Phi(\beta)$. The proof of these facts is easy. Also straightforward is the proof of the following fact, which we shall use in Chapter 6. Let X, Y and $C \in \mathsf{SmProj}(k)$ be pointed varieties with base points e_X, e_Y and e_C respectively and with C a curve. Let $f : A_X \to P_C = J(C)$ and $g : J(C) = A_C \to P_Y$ be homomorphisms, and let $D = \Phi(f)$ and $E = \Phi(g)$ be the corresponding divisor classes from (b), normalized in the above points. Then $E \circ D = \Phi(g \circ f)$.

(d) For later use we add

PROPOSITION A-2.2. *Let $X, Y \in \mathsf{SmProj}(k)$ and D a divisor on $X \times Y$ algebraically equivalent to zero. Then for some integer $m \neq 0$ we have $mD \in \mathsf{CH}^1_{\underline{\underline{\equiv}}}(X \times Y)$.*

Proof: By assumption, $D = \tilde{D}(a) - \tilde{D}(b)$ where $\tilde{D} \in \mathsf{CH}^1(C \times X \times Y)$, and where C is a smooth projective curve and a, b points on C. Fix a point $x_0 \in X(k)$ and let u be the generic point of C; then $\tilde{D}(u) = \mathrm{pr}_{X \times Y}(\tilde{D} \cdot (u \times X \times Y))$ gives a homomorphism

$$\Phi_{\tilde{D}(u)} : \mathrm{Alb}_X \to (\mathrm{Pic}_Y^0)_{\mathrm{red}}, \quad \Phi_{\tilde{D}(u)}(x) = \tilde{D}(u)(x - x_0), \ x \in X.$$

Now $\Phi_{\tilde{D}(u)}$ is defined over the field $K = k(u)$ and hence so is the abelian subvariety $B_u := \left(\mathrm{Ker}\, \Phi_{\tilde{D}(u)} \right)^0$ of Alb_X. Since K is a primary extension of k (i.e. $K \cap \bar{k} = k$) the theorem of Chow [**La**, thm. 5 p. 26] implies that B_u is defined over k and hence does not depend on u. Therefore

$$\left(\mathrm{Ker}\, \Phi_{\tilde{D}(u)} \right)^0 = \left(\mathrm{Ker}\, \Phi_{\tilde{D}(a)} \right)^0 = \left(\mathrm{Ker}\, \Phi_{\tilde{D}(b)} \right)^0.$$

But then there is a positive integer m such that

$$m\Phi_{\tilde{D}(u)} = m\Phi_{\tilde{D}(a)} = m\Phi_{\tilde{D}(b)}.$$

Without loss of generality we can (because of our assertion) assume that $\tilde{D}(a)(x_0) = \tilde{D}(b)(x_0)$ and hence part (b) above and the Claim (previous page) applied to $\Phi_{mD} : A_X \to P_Y$ give $mD \in \mathsf{CH}^1_{\underline{\underline{\equiv}}}(X \times Y)$. $\qquad\square$

Remarks A-2.3. 1) Alternatively, if A and B are two abelian varieties, the group $\mathrm{Hom}_{AV}(A, B)$ is a finitely generated free abelian group [**Mum69**, p. 164, thm. 3], [**La**, p. 184, Cor. 2] and hence $\Phi_{\tilde{D}(u)}$ does not depend on the parameter u.

Very concretely we can see this by the following cohomological argument: without loss of generality we may assume $D(x_0) = 0$. Then D defines a homomorphism $\Phi_D : A_X \to P_Y$, but this is zero as soon as it is zero on the ℓ^ν–torsion points. By [**Mil80**, Cor. 4.18, p.131] the map on ℓ^ν–torsion points is given by a homomorphism

$$H^{2d-1}_{\mathrm{\acute{e}t}}(X_{\bar{k}}, \mathbb{Z}/\ell^\nu\mathbb{Z}) \to H^1_{\mathrm{\acute{e}t}}(Y_{\bar{k}}, \mathbb{Z}/\ell^\nu\mathbb{Z}),$$

which is induced by the Künneth-component $D_{1,1}$ of the class

$$\gamma(D) \in H^2_{\mathrm{\acute{e}t}}(X_{\bar{k}} \times Y_{\bar{k}}, \mathbb{Z}/\ell^\nu\mathbb{Z})$$

of D. This component vanishes since D is homologically equivalent to zero.
2) By Matsusaka's theorem in § A-1 we have

$$\mathsf{CH}^1_{\mathrm{alg}}(X \times Y)_{\mathbb{Q}} = \mathsf{CH}^1_{\mathrm{hom}}(X \times Y)_{\mathbb{Q}} = \mathsf{CH}^1_{\mathrm{num}}(X \times Y)_{\mathbb{Q}}.$$

Hence Proposition A-2.2 also holds for $D \in \mathsf{CH}^1_{\mathrm{hom}}(X \times Y)$ or $D \in \mathsf{CH}^1_{\mathrm{num}}(X \times Y)$.

A-3. Higher Codimension

Here the situation is drastically different as shown for zero cycles by a result of Mumford [**Mum69**] and for cycles of codimension 2 on threefolds by Griffiths [**Griff69**] as we explain now.

A-3.1. Zero-cycles. Using the universality of the Albanese map $\mathrm{alb}_X : X \to \mathrm{Alb}_X$ (see § A-2) we get an intrinsic homomorphism

$$\mathsf{CH}_0^{\mathrm{hom}}(X) = \frac{\{Z \in \mathsf{Z}_0(X) \mid \deg Z = 0\}}{\sim_{\mathrm{rat}}} \xrightarrow{\ \mathrm{alb}_X\ } \mathrm{Alb}_X$$

and its kernel

$$\boxed{T(X) := \mathrm{Ker}\left[\mathsf{CH}_0^{\mathrm{hom}}(X) \xrightarrow{\ \mathrm{alb}_X\ } \mathrm{Alb}_X\right]} \tag{10}$$

is called the *Albanese kernel.* We have the fundamental result:

THEOREM ([**Mum69**]). *Let X be a surface defined over $\mathbb{C}$ and assume that $p_g(X) := \dim H^0(X, \Omega_X^2) \neq 0$, then $T(X)$ is infinite dimensional in the sense that it cannot be parametrized by any algebraic variety.*

This result has been extended by Bloch in 1971:

THEOREM. *Let X be an algebraic surface over an algebraically closed field. Put*

$$\boxed{H_{\mathrm{trans}}^2(X) := H_{\mathrm{\acute{e}t}}^2(X)/H_{\mathrm{alg}}^2(X).}$$

If $H_{\mathrm{trans}}^2(X) \neq 0$ then $T(X) \neq 0$; it is even infinite dimensional in the sense that for no curve C there exists $Z \in \mathsf{CH}^2(C \times X)$ such that the homomorphism $J(C) \to T(X)$, determined by $(x - x_0) \mapsto Z(x) - Z(x_0))$ is onto , where $x_0 \in C$ is some fixed point.

See [**Blo80, Blo-Sri**] for details and further discussion. For instance, in [**Blo80**] one finds a conjectural converse:

CONJECTURE A-3.1 (Bloch Conjecture).

$$H_{\mathrm{trans}}^2(X) = 0 \implies \mathsf{CH}_0^{\mathrm{hom}}(X) \xrightarrow{\sim} \mathrm{Alb}_X .$$

For the torsion points one has:

THEOREM ([**Roi, Blo79, Mil82**]). *Over an algebraically closed field k, one has*

$$\mathsf{CH}_0^{\mathrm{hom}}(X)_{\mathrm{tors}} \xrightarrow{\sim} \mathrm{Alb}_X(k)_{\mathrm{tors}},$$

where the subscript "tors" denotes the torsion.

Indeed, Roitman proved the theorem already around 1970 (for torsion prime to the characteristic) but this was published much later. Bloch gave a different proof of this result, which was extended by Milne to include p-torsion (where p is the characteristic of k).

A-3.2. Algebraic versus homological equivalence. Griffiths proved in 1969 the, at that time, surprising result that algebraic equivalence can be different from homological equivalence in codimension ≥ 2:

THEOREM ([**Griff69**]). *For all $i \geq 1$ set*

$$\mathsf{Z}_\tau^i(X) := \left\{ Z \in \mathsf{Z}^i(X) \mid nZ \sim_{\mathrm{alg}} 0 \text{ for some } n \in \mathbb{Z} \right\}.$$

There are smooth complex varieties X of dimension 3 with

$$\mathsf{Z}_\tau^2(X) \otimes \mathbb{Q} \neq \mathsf{Z}_{\mathrm{hom}}^2(X) \otimes \mathbb{Q}.$$

This has been sharpened as follows:

THEOREM ([**Clem, Voi92**]). *There exist smooth complex varieties X of dimension 3 for which* $\mathsf{Griff}^2(X) \otimes \mathbb{Q}$ *is infinite dimensional. Here*

$$\mathsf{Griff}^i(X) := \mathsf{Z}^i_{\mathrm{hom}}(X)/\mathsf{Z}^i_{\mathrm{alg}}(X) = \mathsf{CH}^i_{\mathrm{hom}}(X)/\mathsf{CH}^i_{\mathrm{alg}}(X).$$

A-3.3. Homological versus numerical equivalence.

THEOREM ([**Liebm**]). *Over $\mathbb{C}$ one has*

(1) *Homological and numerical equivalence coincide, besides in codimension 1 (mentioned already in section A-1) in codimension 2 and in dimension 1.*
(2) *On abelian varieties homological equivalence and numerical equivalence coincide.*

Appendix B: Proof of the Theorem of Voisin–Voevodsky

In this Appendix we shall use the word *correspondence* to mean *correspondence class* and use the notation (11), (12) from Section 2.1.

Our goal is to give proof of the following result.

THEOREM B-1.2 ([**Voe95**], [**Voi96**, p. 267]). *Let X be a smooth projective variety defined over a field k. Then $\mathsf{Z}^i_{\mathrm{alg}}(X)_{\mathbb{Q}} \subset \mathsf{Z}^i_{\otimes}(X)_{\mathbb{Q}}$, i.e., if $Z \sim_{\mathrm{alg}} 0$ there exists $n > 0$ such that $Z^n = 0$ in $\mathsf{CH}^*(X^n)_{\mathbb{Q}}$.*

Proof: The proof proceeds in several steps. Note that since we take cycles with rational coefficients, we can neglect torsion and work over an algebraically closed field k.

Step 1. Reduction to the case of a smooth projective curve.

If $Z \sim_{\mathrm{alg}} 0$ there exist a smooth projective curve C, a correspondence $\Gamma \in \mathsf{Corr}(C, X)$ and two points $a, b \in C(k)$ such that $Z = \Gamma_*(a - b)$. Put $Z_0 = (a - b) \in A_0(C)$. Since $Z = \Gamma_*(Z_0)$, we obtain $Z^n = (\Gamma^n)_*(Z_0^n)$. Hence it suffices to show that $Z_0^n = 0$ on the n–fold product $C \times \cdots \times C$. In fact we shall prove

$$(a - b)^n = 0, \quad \text{for } n \geq g + 1, \quad g = \text{ genus of } C.$$

(Note that we reduced to a very special cycle on the n–fold product of the curve.)

As Z_0^n is invariant under the action of the symmetric group S_n, it belongs to the subgroup $\mathsf{CH}_0(C^n)^{S_n}_{\mathbb{Q}} \cong \mathsf{CH}_0(S^n C)_{\mathbb{Q}}$; cf. [**Ful**, Example 1.7.6].

Step 2. Let C be a smooth projective curve of genus g defined over k, $e \in C(k)$ a base point. On $S^n C$, the n–fold symmetric product of C we use the notation

$$[x_1, \ldots, x_n] = \pi_n(x_1, \ldots, x_n) \in S^n C, \ \pi_n : C^n \to S^n C \text{ the natural surjection.}$$

Moreover, if we have repeated entries, we collect them together and write them as follows

$$[\underbrace{a, \cdots, a}_{k \text{ times}}, \ldots, \underbrace{b, \cdots, b}_{\ell \text{ times}}] = [ka, \cdots, \ell b].$$

Let $\varphi_n : S^n C \to \mathrm{J}(C)$ the map defined by

$$\varphi_n[x_1, \cdots, x_n] = \sum_{i=1}^{n} x_i$$

where the point $x_i \in J(C)$ corresponds to the divisor $(x_i) - (e)$ of degree zero on C. We claim that $(\varphi_n)_* : \mathsf{CH}_0(S^n C) \xrightarrow{\simeq} \mathsf{CH}_0(\mathrm{J}(C))$ for all $n \geq g$.

Proof: First take $n = g$. Then φ_g is a birational morphism and it is known [**Ful**, Ex. 16.1.11, p. 312] that for X smooth and proper over k the group $\mathsf{CH}_0(X)$ is a birational invariant.[8]

Next, turning to $n \geq g$, put $r = n - g$ and consider the embedding $\iota : S^g C \to S^n C$ given by $[x_1, \ldots, x_g] \mapsto [x_1, \ldots, x_g, re]$. Consider the homomorphism

$$(\varphi_g)_* = (\varphi_n)_* \circ \iota_* : \mathsf{CH}_0(S^g C) \to \mathsf{CH}_0(S^n C) \to \mathsf{CH}_0(J(C)).$$

Since $(\varphi_g)_*$ is an isomorphism, ι_* is injective. To show that $(\varphi_n)_*$ is an isomorphism it thus suffices to show that ι_* is also surjective. Let $y \in S^n C$ and put $z = \varphi_n(y) \in J(C)$. There exists $x \in S^g C$ such that $\varphi_g(x) = z$. The points $\iota(x)$ and y belong to the fiber $\varphi_n^{-1}(z)$, but since the fiber is a projective space the points are rationally equivalent and $\iota_*(x) = y$. $\square$

Remark B-1.3. At this point, the result already follows from a theorem of Bloch [**Blo76**]. Namely $\varphi_{g+1,*}(Z_0^{g+1}) = (\varphi_1(a) - \varphi_1(b))^{*(g+1)}$ where $*$ denotes the Pontryagin product of cycles on $J(C)$. This product is zero in $\mathsf{CH}_0(J(C))$ by Bloch's theorem, hence also in $\mathsf{CH}_0(S^{g+1} C)$ by Step 2. We proceed to give a more direct proof. The first author would like to thank U. Jannsen and Srinivas for a very helpful discussion on this topic.

Consider the correspondences

$$\alpha_n : S^n C \to S^{n+1} C, \quad \beta_n : S^{n+1} C \to S^n C$$

defined by

$$\alpha_n([x_1, \ldots, x_n]) = [x_1, \ldots, x_n, e]$$
$$\beta_n([x_1, \ldots, x_{n+1}]) = \sum_{k=0}^{n} \sum_{1 \leq i_1 < \ldots < i_{n-k} \leq n+1} (-1)^k [x_{i_1}, \ldots, x_{i_{n-k}}, ke].$$

Here the summation convention is that for the index $k = n$ there is only one term $(-1)^n [ne]$. Note that the first correspondence is a morphism; the second correspondence appears in [**Kimu-Vi**, Def 1.8]. If no confusion arises we shall drop the subscripts and put $\alpha = \alpha_n$, $\beta = \beta_n$.

Step 3. We have $\beta \circ \alpha = \mathrm{id}_{S^n C}$.

Proof: We have (with the same convention in the summation as used in Step 2):

$$\beta\alpha([x_1, \ldots, x_n]) = \beta([x_1, \ldots, x_n, e])$$
$$= \sum_{k=0}^{n} \sum_{1 \leq i_1 < \ldots < i_{n-k} \leq n} (-1)^k [x_{i_1}, \ldots, x_{i_{n-k}}, ke]$$
$$+ \sum_{k=0}^{n-1} \sum_{1 \leq i_1 < \ldots < i_{n-k-1} \leq n} (-1)^k [x_{i_1}, \ldots, x_{i_{n-k-1}}, (k+1)e].$$

We see that all the terms cancel, except the one with $k = 0$ in the first sum. This gives the desired result. $\square$

---

[8]For a birational $f : X \dashrightarrow Y$ between smooth and proper varieties over k we have ${}^{\mathsf{T}}\Gamma_f \circ \Gamma_f = \mathrm{id}_X +$ a degenerate correspondence (which operates as zero on zero-cycles) and so $f^* \circ f_*$ is the identity on $\mathsf{CH}_0(X)$ and hence f_* is injective and f^* surjective. Interchanging X and Y shows that f_* is an isomorphism

Step 4. The maps $\alpha_* : \mathrm{CH}_0(S^nC) \to \mathrm{CH}_0(S^{n+1}C)$ and $\beta_* : \mathrm{CH}_0(S^{n+1}C) \to \mathrm{CH}_0(S^nC)$ are isomorphisms for all $n \geq g$.

Proof: We have a commutative diagram

$$
\begin{array}{ccc}
S^nC & \xrightarrow{\ \alpha\ } & S^{n+1}C \\
\varphi_n \downarrow & & \downarrow \varphi_{n+1} \\
J(C) & = & J(C).
\end{array}
$$

Hence $(\varphi_n)_* = (\varphi_{n+1})_* \circ \alpha_*$. So, by Step 2, we find that like the homomorphisms $(\varphi_n)_*$ also α_* is an isomorphism for all $n \geq g$. As $\beta_* \circ \alpha_* = \mathrm{id}$ by Step 3, and α_* is an isomorphism, it follows that β_* is an isomorphism. $\qquad\square$

Step 5. The zero-cycle $Z_0 = a - b$ satisfies $\beta_{g*}(Z_0^{g+1}) = 0$. This then finishes the proof since by Step 4 the map β_{g*} is an isomorphism.

Proof: Without loss of generality, we may choose $e = a$ as base point. Applying equation (4) from § 1.2 C with $Z = e - b$ we find that in the Chow group of C^{g+1} the following equality holds

$$
Z_0^{g+1} \;=\; (b-e)^{g+1} = \sum_{k=0}^{g+1}(-1)^k \binom{g+1}{k}(b^{g-k+1} \times e^k).
$$

This cycle being symmetric belongs to the Chow group of $S^{g+1}C$ and there it can be written as

$$
\sum_{k=0}^{g+1}(-1)^k \binom{g+1}{k}[(g-k+1)b, ke] =
$$

$$
= [(g+1)b)] + \sum_{k=1}^{g+1}(-1)^k \binom{g+1}{k}\alpha_{g*}[(g-k+1)b, (k-1)e].
$$

Using the result of Step 3 , namely, $\beta_{g*} \circ \alpha_{g*} = \mathrm{id}$, we obtain

$$
\beta_{g*}(Z_0^{g+1}) \;=\; \beta_{g*}[(g+1)b] + \sum_{k=1}^{g+1}(-1)^k \binom{g+1}{k}[(g-k+1)b, (k-1)e],
$$

and the result follows since by the definition of β_g (see above)

$$
\beta_{g*}[(g+1)b] = \sum_{\ell=0}^{g}(-1)^\ell \binom{g+1}{\ell+1}[(g-\ell)b, \ell e]
$$

which cancels the second term in the right hand side of the above expression. $\qquad\square$

CHAPTER 2

Motives: Construction and First Properties

In this chapter we describe Grothendieck's construction (1964) of motives. Assumptions and notation are as in Chapter 1.

2.1. Correspondences

2.1.1. A *correspondence* from X to Y is a cycle on the product $X \times Y$ (see Chapter 1). However we shall always be working with *correspondence classes* with respect to a suitable adequate equivalence relation. Moreover we shall use rational instead of integral coefficients. Therefore we introduce the following shorthand notation

$$\boxed{\mathsf{Corr}(X,Y) := \mathsf{CH}(X \times Y) \otimes \mathbb{Q} := \mathsf{CH}(X \times Y; \mathbb{Q}).} \tag{11}$$

We call the classes $f \in \mathsf{Corr}(X,Y)$ themselves correspondences and we write $f : X \to Y$ or sometimes $X \vdash Y$ (see [**Ful**, p. 305]). Recall that the transpose is $^{\mathsf{T}}f : Y \vdash X$.

If we work with another adequate equivalence relation $\sim$, then we write as in § 1.2

$$f \in \mathsf{Corr}_\sim(X,Y) := C_\sim(X \times Y; \mathbb{Q}).$$

2.1.2. Composition of correspondences. For $f \in \mathsf{Corr}_\sim(X,Y)$ and $g \in \mathsf{Corr}_\sim(Y,Z)$ we define the composition $g{\circ}f \in \mathsf{Corr}_\sim(X,Z)$ by the formula

$$g{\circ}f := \mathrm{pr}_{XZ}\left\{(f \times Z) \cdot (X \times g)\right\}$$

where the intersection is in $C_\sim(X \times Y \times Z)$ (always defined!). Composition gives a map

$$\mathsf{Corr}_\sim(X,Y) \times \mathsf{Corr}_\sim(Y,Z)) \to \mathsf{Corr}_\sim(X,Z)$$

and when $X = Y = Z$ this makes $\mathsf{Corr}_\sim(X,X)$ a ring (in general non-commutative!).

DEFINITION 2.1.1. A *projector* for X is an element $p \in \mathsf{Corr}_\sim(X,X)$ for which $p{\circ}p = p$.

2.1.3. Degree of a correspondence. Let X_d and Y be (smooth and projective) varieties (also recall our convention from Chapter 1 that X_d means that X is irreducible of dimension d). Put (see Chapter 1)

$$\boxed{\mathsf{Corr}_\sim^r(X_d,Y) := C_\sim^{d+r}(X \times Y; \mathbb{Q}) \quad \text{(degree } r \text{ correspondences).}} \tag{12}$$

Note that $\mathsf{Corr}_\sim^0(X,X) \subset \mathsf{Corr}_\sim(X,X)$ is a *subring* and that if p is a projector, then p has degree 0. Finally, observe that if $\phi : X_d \to Y_e$ is a morphism in the usual sense, the graphs $\Gamma_f \subset X \times Y$ define

$$
\begin{aligned}
\phi_* &:= \Gamma_\phi \in \mathsf{Corr}_\sim^{e-d}(X,Y) \\
\phi^* &:= {}^{\mathsf{T}}\Gamma_\phi \in \mathsf{Corr}_\sim^0(Y,X).
\end{aligned}
$$

23

2.1.4. Operation on cycle groups and cohomology. Any correspondence $f \in \mathsf{Corr}^r_\sim(X, Y)$ induces homomorphisms (in fact a $\mathbb{Q}$-linear maps of vector spaces)

$$f = f_* : C^i_\sim(X; \mathbb{Q}) \quad \to \quad C^{i+r}_\sim(Y; \mathbb{Q})$$
$$Z \quad \mapsto \quad f_*(Z) := \mathrm{pr}_Y \left\{ f \cdot (\mathrm{pr}_X)^*(Z) \right\}.$$

In other words:

$$\boxed{\text{Degree } r \text{ correspondences send codim. } i\text{--cycles to codim. } (i+r)\text{--cycles.}}$$

Note that $f \in \mathsf{Corr}^0_\sim(X, Y)$ respects the degree. Similarly, if $f \in \mathsf{Corr}^r_\sim(X, Y)$ for an equivalence relation with $\mathsf{Z}_\sim(-) \subseteq \mathsf{Z}_{\mathrm{hom}}(-)$ we have an operation on a Weil cohomology

$$f_* : H^i(X) \quad \to \quad H^{i+2r}(Y)$$
$$\alpha \quad \mapsto \quad f_*(\alpha) := \mathrm{pr}_Y \left\{ \gamma_{X \times Y}(f) \cup \mathrm{pr}_X^*(\alpha) \right\}.$$

Remarks 2.1.2.   (1) If $\phi : X \to Y$ is a morphism of varieties then Γ_ϕ is the usual operation ϕ_* and $^\mathsf{T}\Gamma_\phi$ the usual ϕ^*.

   (2) The above operation on cohomology indicates the significance of the conjecture $D(X)$ stating that homological and numerical equivalence coincide, because a correspondence $f \in \mathsf{Corr}_{\mathrm{num}}(X, Y)$ operates on cohomology only if this conjecture is true.

2.1.5. We often use the important

LEMMA 2.1.3 (Lieberman's Lemma. [**Ful**, p. 306], [**Klei70**, p. 73]). *Let $f \in \mathsf{Corr}_\sim(X, Y), \alpha \in \mathsf{Corr}_\sim(X, X'), \beta \in \mathsf{Corr}_\sim(Y, Y')$, then $(\alpha \times \beta)_*(f) = \beta \circ f \circ {}^\mathsf{T}\alpha$ (and similar statements in "the other directions"). Note that the left hand side is the operation of $\alpha \times \beta$ on f while the right hand side is the composition of correspondences.*

$$
\begin{array}{ccc}
X & \longmapsto\!\!f\!\!\longmapsto & Y \\
{\scriptstyle\alpha}\big\uparrow & & \big\uparrow{\scriptstyle\beta} \\
X' & \vdash - - & Y'
\end{array}
$$

Proof: By definition

$$(\alpha \times \beta)_*(f) = (p^{XX'YY'}_{X'Y'})_* ((\alpha \times \beta) \cdot (p^{XX'YY'}_{XY})^*(f)),$$

where $p^{XX'YY'}_{X'Y'}$ (resp. $p^{XX'YY'}_{XY}$) denotes the projection from $X \times X' \times Y \times Y'$ to $X' \times Y'$ (resp. $X \times Y$). Using the isomorphism

$$X \times X' \times Y \times Y' \cong X' \times X \times Y \times Y'$$

we can rewrite this expression as

$$
\begin{aligned}
(\alpha \times \beta)_*(f) &= (p^{X'XYY'}_{X'Y'})_* (({}^\mathsf{T}\alpha \times \beta) \cdot (X' \times f \times Y')) \\
&= (p^{X'XYY'}_{X'Y'})_* (({}^\mathsf{T}\alpha \times Y \times Y') \cdot (X' \times X \times \beta) \cdot (X' \times f \times Y')).
\end{aligned}
$$

We now rewrite the right hand side of this expression using the commutative diagram

$$
\begin{array}{ccc}
X' \times X \times Y \times Y' & \xrightarrow{\ p\ } & X' \times Y \times Y' \\
{\scriptstyle p^{X'XYY'}_{X'Y'}}\big\downarrow & \swarrow{\scriptstyle p^{X'YY'}_{X'Y'}} & \\
X' \times Y' & &
\end{array}
$$

where $p = p_{X'YY'}^{X'XYY'} = p_{X'Y}^{X'XY} \times \mathrm{id}_{Y'}$. We obtain

$$
\begin{aligned}
(\alpha \times \beta)_*(f) &= (p_{X'Y'}^{X'YY'})_*[p_*\{({}^{\mathsf{T}}\alpha \times Y \times Y') \cdot (X' \times X \times \beta) \cdot (X' \times f \times Y')\}] \\
&= (p_{X'Y'}^{X'YY'})_*[p_*\{({}^{\mathsf{T}}\alpha \times Y \times Y') \cdot (X' \times f \times Y') \cdot p^*(X' \times \beta)\}]
\end{aligned}
$$

By the projection formula we have

$$
p_*\{({}^{\mathsf{T}}\alpha \times Y \times Y') \cdot (X' \times f \times Y')) \cdot p^*(X' \times \beta)\} = p_*\{({}^{\mathsf{T}}\alpha \times Y \times Y') \cdot (X' \times f \times Y')\} \cdot (X' \times \beta).
$$

Since $p = p_{X'Y}^{X'XY} \times \mathrm{id}_{Y'}$, we have

$$
p_*\{({}^{\mathsf{T}}\alpha \times Y \times Y') \cdot (X' \times f \times Y')\} = (f \circ {}^{\mathsf{T}}\alpha) \times Y'.
$$

Substituting this in the right hand side of the above expression, we obtain

$$
(p_{X'Y'}^{X'YY'})_*[\{(f \circ {}^{\mathsf{T}}\alpha) \times Y'\} \cdot (X' \times \beta)] = \beta \circ f \circ {}^{\mathsf{T}}\alpha.
$$

So finally $(\alpha \times \beta)_*(f) = \beta \circ f \circ {}^{\mathsf{T}}\alpha$.

$\square$

2.2. (Pure) Motives

For Grothendieck's motivation we refer to the introduction. We want to stress that the **construction of motives is unconditional**, i.e. free from conjectures, and in fact is very simple. However, for the deeper properties we need the standard conjectures, see Chapter 3.

2.2.1. Grothendieck's Construction.
Fix an adequate equivalence relation $\sim$. The construction of the category $\mathsf{Mot}_\sim(k)$ of motives with respect to $\sim$ proceeds in several steps

$$
\mathsf{SmProj}(k)^{\mathrm{opp}} \to C_\sim\mathsf{SmProj}(k) \to \mathsf{Mot}_\sim^{\mathrm{eff}}(k) \hookrightarrow \mathsf{Mot}_\sim(k).
$$

<u>Step 1</u>. Recall that $\mathsf{SmProj}(k)$ is the category of smooth projective varieties defined over k and morphisms are the usual morphisms between varieties. If $\mathfrak{C}$ is any category, $\mathfrak{C}^{\mathrm{opp}}$ denotes the category with the same objects as $\mathfrak{C}$ but with the morphisms reversed.

<u>Step 2</u>. $C_\sim\mathsf{SmProj}(k)$ has the same objects (smooth projective varieties) but the morphisms are the degree zero correspondences and the composition is the composition of correspondences. Note that $C_\sim\mathsf{SmProj}(k)$ is an additive category.

<u>Step 3</u>. $\mathsf{Mot}_\sim^{\mathrm{eff}}(k)$ (category of effective motives). The objects are pairs (X, p) with $X \in \mathsf{SmProj}(k)$, p a projector, and where the morphisms $(X, p) \to (Y, q)$ are of the form $f = q \circ f' \circ p$ with f' a degree 0 correspondence:

$$
\mathrm{Hom}_{\mathsf{Mot}_\sim^{\mathrm{eff}}}((X, p), (Y, q)) = q \circ \mathrm{Corr}_\sim^0(X, Y) \circ p.
$$

Composition comes from composition of correspondences.

Remarks. (1) Note that $f = q \circ f' \circ p$ implies $f = q \circ f = f \circ p$: all subdiagrams in

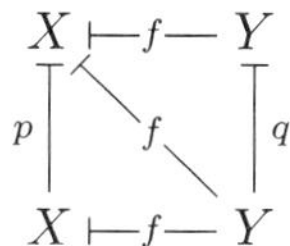

commute.

(2) $\mathsf{Mot}^{\mathrm{eff}}_\sim(k) = (C_\sim\mathsf{SmProj}(k))^\natural$, the pseudo-abelian (or Karoubian) completion. By definition, the *pseudo-abelian completion* $\mathfrak{C}^\natural$ of an additive category $\mathfrak{C}$ is such that the projectors from $\mathfrak{C}$ have images and kernels in the category $\mathfrak{C}^\natural$; furthermore, the functor $A \mapsto (A, \mathrm{id})$ (from $\mathfrak{C}$ to its completion) is universal for functors from $\mathfrak{C}$ to pseudo-abelian categories.

One can show that projectors in $\mathfrak{C}^\natural$ have images and kernels in $\mathfrak{C}$. For instance if $M = (X, p, 0) \in \mathsf{Mot}_\sim(k)$ then a projector q of M is an element $q = p{\circ}q'{\circ}p$ with $q' \in \mathsf{Corr}^0_\sim(X, X)$ such that $q{\circ}q = q$ and $N = (X, q, 0)$ is the image of q. Note that $q = p{\circ}q = q{\circ}p$, i.e., N is a *constituent* of M. So is $N' = \mathrm{Ker}(q) = (X, p - q, 0)$, and we have $M = N \oplus N'$. Hence in $\mathfrak{C}^\natural$ we can split off kernels and images as direct summands.

Step 4 Final step: the motives. Objects are triples (X, p, m) with $X \in \mathsf{SmProj}(k)$, p a projector, $m \in \mathbb{Z}$. The morphisms are as follows

$$\mathrm{Hom}_{\mathsf{Mot}_\sim}((X, p, m), (Y, q, n)) = q{\circ}\mathsf{Corr}^{n-m}_\sim(X, Y){\circ}p$$

and composition of morphisms comes from composition of correspondences. Clearly we have a faithful full embedding $\mathsf{Mot}^{\mathrm{eff}}_\sim \hookrightarrow \mathsf{Mot}_\sim$. The enlargement is related to Tate twists in cohomology as we shall see below. $\qquad\square$

Two equivalence relations are particularly important:

(a) *Rational equivalence* (the finest equivalence relation). We write sometimes

$$\mathsf{CHM} = \mathsf{Mot}_{\mathrm{rat}}$$

for these so-called *Chow motives*. This category is additive and pseudo-abelian, but not abelian. See for example [**Scholl**, Cor. 3.5, p. 173].

(b) *Numerical equivalence* (the coarsest equivalence relation). We write sometimes

$$\mathsf{NM} = \mathsf{Mot}_{\mathrm{num}}$$

for the category of motives modulo numerical equivalence.[1] This category turns out to be abelian and semi-simple (Jannsen's theorem, see Chapter 3).

Returning to the general case, we have a contravariant functor

$$
\boxed{
\begin{array}{rcc}
h_\sim : \mathsf{SmProj}(k) & \to & \mathsf{Mot}_\sim(k) \\
X & \mapsto & h_\sim(X) = (X, \Delta_X, 0) \\
f : X \to Y & \mapsto & h_\sim(f) = {}^{\mathsf{T}}\Gamma_f : h_\sim(Y) \to h_\sim(X)
\end{array}
}
$$

that sends a smooth, projective variety X to its motive $h_\sim(X) = (X, \Delta_X, 0)$. In the special case $\sim\ =$ rational equivalence, one writes

$$\mathsf{ch}(X) := h_{\mathrm{rat}}(X)$$

for the functor that sends X to its Chow motive $h_{\mathrm{rat}}(X)$.

Remarks. (1) This construction gives the so-called contravariant pure motives adapted to cohomology, i.e., aimed at giving a universal cohomology theory. The category $\mathsf{Mot}^{\mathrm{OPP}}_\sim$ of covariant pure motives is aimed

---

[1] This is sometimes called the category of *Grothendieck motives*, since Grothendieck placed special emphasis on this category; cf. [**Groth69a**, p. 198]

towards homology; it is related to the covariant approach of Voevodsky, as will be explained in Chapter 9.

(2) The previous construction shows that every variety can be seen as a motive. Intuitively, a (general) motive $M = (X, p, 0)$ should be seen as a "piece" of X that is "responsible" for a certain part of the geometric and (or) arithmetic properties of X, see for instance example ii) below. To be precise, $(X, p, 0)$ is a direct summand of $h_\sim(X)$; and by [**Scholl**, Prop. 1.12] for $m > 0$, (X, p, m) is a direct summand of

$$h_\sim(X \times \underbrace{\mathbb{P}^1 \times \cdots \times \mathbb{P}^1}_{m}).$$

Note that the same motive can be realized as a "piece" of different varieties. See § 2.4 (v) below.

2.2.2. Categorical Aspects. A morphism $f : (X, p) \to (Y, q)$ of (effective) motives is by definition of the form $q \circ f' \circ p$, where f' is a correspondence of degree zero from X to Y. This can be rephrased as follows. Consider the subgroup

$$\mathrm{Hom}^*\left((X, p), (Y, q)\right) := \left\{ f \in \mathsf{Corr}^0_\sim(X, Y) \mid f \circ p = q \circ f \right\}$$

of the group $\mathsf{Corr}^0_\sim(X, Y)$. On this subgroup we define an equivalence relation by declaring $f \approx 0$ if $q \circ f = f \circ p = 0$. In particular, in $\mathrm{Hom}^*\left((X, p), (X, p)\right)$ we have $p \approx \mathrm{id}_X$. Then $f \approx q \circ f \approx q \circ f \circ p$. Let $[f]$ be the equivalence class of f in $\mathrm{Hom}^*\left((X, p), (Y, q)\right)$. Then $[f] = [q][f][p] = [f][p] = [q][f]$ and one shows easily that sending $f = q \circ f' \circ p$ to $[f] = [q \circ f' \circ p] = [f']$ induces an isomorphism [2]

$$\mathrm{Hom}_{\mathsf{Mot}_\sim}\left((X, p), (Y, q)\right) \cong \mathrm{Hom}^*\left((X, p), (Y, q)\right) / \approx.$$

Two effective motives $M = (X, p)$ and $N = (Y, q)$ are isomorphic if one has two degree zero correspondences $f' : X \to Y$ and $g' : Y \to X$ such that for the corresponding morphisms $f = q \circ f' \circ p$ and $g = p \circ g' \circ q$ one has $g \circ f = p = \mathrm{id}_M$ and $f \circ g = q = \mathrm{id}_N$. In terms of f' and g' this means $p \circ g' \circ q \circ f' \circ p = p$ and $q \circ f' \circ p \circ g' \circ q = q$. More generally, for two motives $M = (X, p, m)$ and $N = (Y, q, n)$ to be isomorphic, one demands that instead the correspondences f', g' have degree $n - m$ and $m - n$ respectively, and such that the same relations as before hold.

2.3. Examples

For the moment we only give rather trivial examples.

(i) The motive of a point

$$\boxed{\mathbf{1} := (\operatorname{Spec} k, \mathrm{id}, 0) = \text{``}h_\sim(\text{point})\text{''}.}$$

(ii) Assume $X_d(k) \neq \varnothing$ (can always be achieved by enlarging k if necessary)[3] and pick $e \in X(k)$. Define

$$p_0(X) := e \times X, \quad p_{2d}(X) = X \times e.$$

---

[2]The right-hand side was used as the original definition of the morphisms in $\mathsf{Mot}_\sim(k)$; see [**Manin**, section 5, p. 453].

[3]In case $X(k) = \varnothing$ and if we insist on keeping k as ground field we can take a k-rational positive 0–cycle $\mathfrak{A}$ of degree n and take $p_0(X) = \frac{1}{n}\mathfrak{A} \times X$ and $p_{2d}(X) = \frac{1}{n}X \times \mathfrak{A}$.

These define two projectors which are orthogonal in the sense that $p_0{\circ}p_{2d} = p_{2d}{\circ}p_0 = 0$. This then defines the two motives

$$h^0_\sim(X) = (X, p_0(X), 0), \quad h^{2d}_\sim(X) = (X, p_{2d}(X), 0).$$

We leave it as an exercise to see that $h^0(X) \simeq \mathbf{1}$ and that for all varieties X of dimension d the motives $h^{2d}(X)$ are mutually isomorphic. (Indication [**Scholl**] : use the structural morphism $X \to \operatorname{Spec} k$ and the morphism $\operatorname{Spec} k \to X$ given by the point e).

(iii) *Direct sums of motives.* Let $M = (X, p, m)$ and $N = (Y, q, m)$. Then one can define a motive $M \oplus N$. Let us only give the definition in case $m = n$:

$$M \oplus N = (X \sqcup Y, p \sqcup q, m)$$

and refer to [**Scholl**, 1.14, p. 169] for the general case. Here $\sqcup$ denotes the disjoint union.

As an application, for any $X_d \in \mathsf{SmProj}(k)$, the correspondence $p^+(X) := \Delta - p_0(X) - p_{2d}(X)$ is a projector (since p_0 and p_{2d} are orthogonal) and if we put $h^+_\sim(X) := (X, p^+(X), 0)$ there is a direct sum decomposition

$$h_\sim(X) = h^0_\sim(X) \oplus h^+_\sim(X) \oplus h^{2d}_\sim(X).$$

(iv) *Lefschetz motive and Tate motive* The *Lefschetz motive* is

$$\boxed{\mathbf{L}_\sim := (\mathbb{P}^1, \mathbb{P}^1 \times e, 0) = h^2_\sim(\mathbb{P}^1)}$$

and the Tate motive is

$$\boxed{\mathbf{T}_\sim := (\operatorname{Spec} k, \operatorname{id}, 1).}$$

As an exercise, show that $\mathbf{L}_\sim \simeq (\operatorname{Spec} k, \operatorname{id}, -1)$ and $h^{2d}_\sim(X) \simeq (\operatorname{Spec} k, \operatorname{id}, -d)$.

(v) If $M = (X, p, 0)$ is a motive, then $1 - p$ is also a projector and $h_\sim(X) = (X, p, 0) \oplus (X, 1 - p, 0)$.

(vi) If $\phi : X_d \to Y_d$ is a generically finite morphism of degree m, then

$$\phi_*{\circ}\phi^* = \Gamma_\phi{\circ}{}^{\mathsf{T}}\Gamma_\phi = m \operatorname{id}_Y$$

and $p := \frac{1}{m}\phi^*{\circ}\phi_*$ is a projector on X. In fact we have $(X, p, 0) \simeq h_\sim(Y)$ (check this!) and hence

$$h_\sim(X) \simeq (X, p, 0) \oplus (X, 1 - p, 0) \simeq h_\sim(Y) \oplus (X, 1 - p, 0).$$

(vii) More generally, let $M = (X, p, 0)$ and $N = (Y, q, 0)$ and assume that we have two morphisms $\alpha : N \to M$ and $\beta : M \to N$ such that $\beta{\circ}\alpha = \operatorname{id}_N = q$. We leave it to the reader to check that $p' = \alpha{\circ}\beta$ is a projector on X and on M, that $N \simeq (X, p', 0)$ and hence that

$$M \simeq N \oplus (X, p - p', 0).$$

2.4. Further Remarks and Properties

(i) $\mathsf{Mot}_\sim(k)$ is an additive, pseudo-abelian category, also sometimes called $\mathbb{Q}$-linear, pseudo-abelian, since the set $\operatorname{Hom}_{\mathsf{Mot}_\sim}(M, N)$ is a $\mathbb{Q}$-vector space.

(ii) The category $\mathsf{Mot}_\sim(k)$ is a tensor category:

$$\mathsf{Mot}_\sim(k) \times \mathsf{Mot}_\sim(k) \xrightarrow{\otimes} \mathsf{Mot}_\sim(k)$$
$$(X, p, m) \otimes (Y, q, n) := (X \times Y, p \times q, m + n).$$

(iii) *Tate twists.* The Tate and Lefschetz motives are inverses:

$$\mathbf{L}_\sim \otimes \mathbf{T}_\sim \simeq \mathbf{1}.$$

Moreover, $h_\sim^{2d}(X_d) \simeq \mathbf{L}_\sim^{\otimes d}$ and

$$(X, p, m) \simeq (X, p, 0)(m) := (X, p, 0) \otimes \mathbf{T}_\sim^{\otimes m}.$$

(iv) The *duality operator*

$$\mathsf{Mot}_\sim(k)^{\mathrm{opp}} \xrightarrow{\ \mathsf{D}\ } \mathsf{Mot}_\sim(k)$$
$$M = (X_d, p, m) \ \mapsto \ \mathsf{D}(M) := (X_d, {}^{\mathsf{T}}p, d - m)$$

is an involution: $\mathsf{D}(\mathsf{D}(M)) = M$. The functor

$$\left. \begin{array}{rl} h_*^\sim \ : & \mathsf{D}\circ h_\sim : \mathsf{SmProj}(k) \to \mathsf{Mot}_\sim(k)^{\mathrm{opp}} \\ & X \mapsto (X, \mathrm{id}, d), \quad d = \dim X \end{array} \right\} \tag{13}$$

is a *covariant* functor, which is adapted to homology rather than cohomology.

(v) The functor

$$h_{\mathrm{rat}} : \mathsf{SmProj}(k) \to \mathsf{Mot}_{\mathrm{rat}}^{\mathrm{eff}}(k)$$

is *not* conservative (injective on objects): We shall see below (Examples 2.8.1(1)) that any two $\mathbb{P}^m$-bundles coming from vector bundles on the same smooth projective variety have isomorphic motives.

2.5. Chow Groups and Cohomology Groups of Motives

(1) **Chow groups.** Recall (Chapter 1) that $\mathsf{CH}^i(X)_\mathbb{Q} = C_{\mathrm{rat}}^i(X; \mathbb{Q}) = \mathsf{CH}^i(X) \otimes \mathbb{Q}$. For any projector $p : X \to X$, for all i there are induced maps $p_* : \mathsf{CH}^i(X)_\mathbb{Q} \to \mathsf{CH}^i(X)_\mathbb{Q}$ and for the motive $M = (X, p, m)$ one defines

$$\boxed{\ \mathsf{CH}^i(M) := \operatorname{Im} p_* \subset \mathsf{CH}^{i+m}(X)_\mathbb{Q} \quad \text{(the } i\text{--th Chow group of } M\text{).}\ }^4$$

PROPOSITION 2.5.1. *One has*

$$\mathsf{CH}^i(M) \simeq \operatorname{Hom}_{\mathsf{Mot}_{\mathrm{rat}}(k)}(\mathbf{L}^{\otimes i}, M).$$

Proof: If $M = (X, p, m)$ then

$$\operatorname{Hom}_{\mathsf{Mot}(\mathrm{rat})k}(\mathbf{L}^{\otimes i}, M) = \{ f = p{\circ}\Gamma \mid \Gamma \in \mathsf{CH}^{i+m}(X) \},$$

and by Lieberman's lemma $p{\circ}\Gamma = (\mathrm{id}_{\mathrm{Spec}\,k} \times p)_*(\Gamma) = \operatorname{Im} p_*$ in $\mathsf{CH}^{i+m}(X)$. $\square$

(2) **Cycle groups.** In the same way, for any $M \in \mathsf{Mot}_\sim$ we can define $C_\sim^i(M)$.

Remark. The formula of Proposition 2.5.1 holds for every adequate equivalence relation: $C_\sim^i(M) = \operatorname{Hom}_{\mathsf{Mot}_\sim(k)}(\mathbf{L}_\sim^{\otimes i}, M)$.

---

[4] Of course it is actually a $\mathbb{Q}$-vector space.

(3) **Cohomology groups.** Let $\sim$ be an equivalence relation finer than (or equal to) homological equivalence, i.e.

$$Z_\sim(-) \subseteq Z_{\mathrm{hom}}(-).$$

Then for the motive $M = (X, p, m)$ the induced maps $p_* : H^i(X) \to H^i(X)$ are the ingredients to define the cohomology groups

$$\boxed{H^i(M) := \operatorname{Im} p_* \subset H^{i+2m}(X).}$$

Note that as special cases we have $H((X, p_0, 0)) = H^0(X)$ and $H((X, p_{2d}, 0)) = H^{2d}(X)$.

In particular, in $\mathsf{Mot}_{\mathrm{rat}}, \mathsf{Mot}_{\mathrm{alg}}, \mathsf{Mot}_\otimes$ and $\mathsf{Mot}_{\mathrm{hom}}$ the motives have cohomology groups, but not in $\mathsf{Mot}_{\mathrm{num}}$, unless the standard conjecture $D(-)$ (homological equivalence = numerical equivalence) is true.

2.6. Relations Between the Various Categories of Motives

Let us summarize the relations between the various notions of motives by way of a commutative diagram of functors:

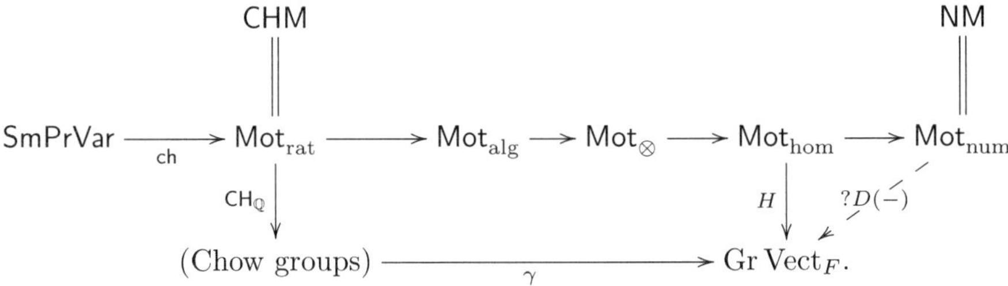

Here the subscript "hom" stands for a Weil cohomology theory with coefficients in a field F of characteristic zero and γ is the cycle map. Again we stress that $\mathsf{Mot}_{\mathrm{hom}}$ depends on the choice of cohomology theory, but in characteristic zero we get the same for each of the classical Weil cohomology theories (as treated in Examples 1.2.14), thanks to the comparison theorems for Betti, De Rham and étale cohomology. Also note that étale cohomology is always on $X \times_k \bar{k}$.

Again, note the fundamental importance of conjecture $D(-)$.

2.7. Motives of Curves

In this section we - finally - encounter a non-trivial example of a motive. Let C be a smooth, projective curve defined over a field k, which we assume for simplicity to be algebraically closed. In this section we shall work in the category of *integral Chow motives*.

DEFINITION 2.7.1. The category $\mathsf{CHM}_\mathbb{Z}(k)$ of integral Chow motives is the category of motives constructed using Chow groups with integer coefficients (not $\mathbb{Q}$-coefficients). The functor that associates to a smooth, projective variety over k its integral Chow motive will be denoted $\mathsf{ch}_\mathbb{Z}(-)$.

Fix a base point $e \in C(k)$. The projectors $p_0(C) = e \times C$ and $p_2(C) = C \times e$ have been introduced before, as well as $p^+(C) = \Delta(C) - p_0(C) - p_2(C)$. We write now

$p_1(C) := p^+(C)$. So we have a so called (integral) *Chow-Künneth decomposition*
(See Chapter 6)

$$\mathsf{ch}_{\mathbb{Z}}(C) = \mathsf{ch}_{\mathbb{Z}}^0(C) \oplus \mathsf{ch}_{\mathbb{Z}}^1(C)) \oplus \mathsf{ch}_{\mathbb{Z}}^2(C), \quad \mathsf{ch}_{\mathbb{Z}}^i(C) := (C, p_i(C), 0). \qquad (14)$$

Fixing a Weil cohomology theory, for the cohomology of the rational motives we
have

$$H(\mathsf{ch}^0(C)) = H^0(C),\, H(\mathsf{ch}^2(C)) = H^2(C), \implies H(\mathsf{ch}^1(C)) = H^1(C).$$

However $\mathsf{ch}^1(C)$ carries much finer information. Indeed this motive is closely related
to the Jacobian variety, $J(C)$ of C. This becomes clear from the following theorem
of Grothendieck (circa 1965), as explained in [**Dem**] which in fact is based on
"classical" results for abelian varieties.

THEOREM 2.7.2. *With the previous notation and assumptions we have:*

(a) *The only non-trivial Chow group of $\mathsf{ch}_{\mathbb{Z}}^1(C)$ is*

$$\mathsf{CH}^1(\mathsf{ch}_{\mathbb{Z}}^1(C)) = J(C)(k) = \mathsf{CH}_{\mathrm{alg}}^1(C),$$

(b) *Let C, C' be smooth projective curves, then*

$$\mathrm{Hom}_{\mathsf{CHM}_{\mathbb{Z}}}(\mathsf{ch}_{\mathbb{Z}}^1(C), \mathsf{ch}_{\mathbb{Z}}^1(C')) \simeq \mathrm{Hom}_{\mathrm{AV}}(J(C), J(C')). \qquad (15)$$

(c) *In the category of rational Chow motives $\mathsf{Mot}_{\mathrm{rat}} = \mathsf{CHM}_{\mathbb{Q}}$ we have that
the full subcategory M' whose objects are direct factors of $\mathsf{ch}^1(C)$, C some
(possibly reducible) smooth projective curve, is equivalent to the category
of abelian varieties up to isogeny.* [5]

Proof: (a) It is clear that $\mathsf{ch}_{\mathbb{Z}}^1(C)$ only has a non-trivial first Chow group i.e. for
divisors. If $\mathfrak{a} \in \mathsf{CH}^1(C)$, we have $p_1(\mathfrak{a}) = \mathfrak{a} - \deg \mathfrak{a} \cdot e$, a divisor of degree 0 and
hence $\mathrm{Im}(p_1) \subset \mathsf{CH}_{\mathrm{alg}}^1(C)$ (numerical and algebraic equivalence coincide on curves).
But on $\mathsf{CH}_{\mathrm{alg}}^1(C)$ the projector p_1 is an isomorphism.
(b) This is essentially a reformulation of a theorem of Weil for Jacobians (Appendix A-2.1) as we now explain. From the definition of morphisms in $\mathsf{CHM}_{\mathbb{Z}}$ we obtain
a homomorphism

$$\psi_1 : \mathsf{CH}^1(C \times C') \twoheadrightarrow \mathrm{Hom}_{\mathsf{CHM}_{\mathbb{Z}}}(\mathsf{ch}^1(C), \mathsf{ch}^1(C'))$$

upon setting $\psi_1(T) = p_1(C') \circ T \circ p_1(C)$, $T \in \mathsf{CH}^1(C \times C')$. It is easy to see that
the kernel is $\mathsf{CH}_{\equiv}^1(C \times C')$ and therefore we get the isomorphism ψ in the diagram
below

$$\mathsf{CH}^1(C \times C') \xrightarrow{\;\;\psi_1\;\;} \mathrm{Hom}_{\mathsf{CHM}_{\mathbb{Z}}}(\mathsf{ch}^1(C), \mathsf{ch}^1(C'))$$
$$\downarrow \qquad\qquad\qquad \nearrow \;\psi$$
$$\mathsf{CH}^1(C \times C')/\mathsf{CH}_{\equiv}^1(C \times C')$$

Combining this with the theorem of Weil (Appendix A-2.1) we get the required
isomorphism

$$\mathrm{Hom}_{\mathsf{CHM}_{\mathbb{Z}}}(\mathsf{ch}^1(C), \mathsf{ch}^1(C')) \xrightarrow{\;\sim\;} \mathrm{Hom}_{\mathrm{AV}}(J(C), J(C')).$$

---

[5]Objects in this category are the abelian varieties (over a fixed field), and the morphisms are
given by $\mathrm{Hom}_{\mathrm{isog}}(A, B) = \mathrm{Hom}_{\mathrm{AV}}(A, B) \otimes \mathbb{Q}$

Consider the full subcategory $\mathsf{M}''_{\mathbb{Z}}$ of $\mathsf{CHM}_{\mathbb{Z}}$ of motives isomorphic to $\mathsf{ch}^1_{\mathbb{Z}}(C)$ for some smooth projective curve C, and let

$$F : \mathsf{M}''_{\mathbb{Z}} \to \{\text{category of Jacobians of curves}\}$$

be the functor defined by $F(\mathsf{ch}^1_{\mathbb{Z}}(C)) = J(C)$. Since F is clearly essentially surjective and is fully faithful by (15), F is an equivalence of categories. Passing to $\mathbb{Q}$-coefficients in the correspondences and taking $\mathrm{Hom}(J(C), J(C')) \otimes \mathbb{Q}$ we have an equivalence

$$F_{\mathbb{Q}} : \mathsf{M}''_{\mathbb{Q}} \xrightarrow{\sim} \{\text{category of Jacobians of curves}\} \otimes \mathbb{Q}.$$

c) For this, Grothendieck again uses known theorems of abelian varieties, in particular the Poincaré reducibility theorem [**Mum74**, Chap. IV §19, Thm. 1]. Next, let A be an abelian variety. Then there exists an algebraic curve C and a surjection $J(C) \twoheadrightarrow A$. By the reducibility theorem of Poincaré, we have $J(C) = A_1 + A_2$ with A_1 and A_2 two abelian subvarieties of $J(C)$ with $A_1 \cap A_2$ a finite group and A_1 isogenous to A. Hence up to isogeny $J(C) = A_1 \oplus A_2$. In other words: the category of abelian varieties up to isogeny is exactly the pseudo-abelian completion of the category of Jacobians of curves (recall that to make the pseudo-abelian completion one has to add images and kernels of projectors). On the other hand, in the statement c) we clearly have[6] $(\mathsf{M}'')^{\natural}_{\mathbb{Q}} = \mathsf{M}'$. Therefore, using the universality of the pseudo-abelian completion, one obtains the following commutative diagram of functors, whence the desired equivalence $F^{\natural}$:

$$
\begin{array}{ccc}
\mathsf{M}''_{\mathbb{Q}} & \xrightarrow{\ \sim\ ,\ F_{\mathbb{Q}}\ } & \{\text{category of Jacobians of curves}\} \otimes \mathbb{Q} \\
\Big\downarrow{\natural} & & \Big\downarrow{\natural} \\
\mathsf{M}' = \mathsf{M}''^{\natural}_{\mathbb{Q}} & \xrightarrow{\ \sim\ ,\ F^{\natural}\ } & \{\text{category of abelian varieties up to isogeny}\} .
\end{array}
$$

$\square$

Remarks 2.7.3.   (a) From (c) it follows that $\mathsf{Mot}_{\mathrm{rat}}$ contains the category of abelian varieties up to isogeny as a full subcategory! It seems likely that this insight must have been also one of the motivations that led Grothendieck to his theory of motives.

(b) Parts (b) and (c) of the above theorem also hold in the categories $\mathsf{Mot}_{\mathrm{alg}}$, $\mathsf{Mot}_{\otimes}$, $\mathsf{Mot}_{\mathrm{hom}}$ and $\mathsf{Mot}_{\mathrm{num}}$ but with $\mathbb{Q}$-coefficients. Because of the nature of the proof, it suffices to see this for part (b). Take first $\mathsf{Mot}_{\mathrm{alg}}$. We have seen (Prop. A-2.2) that

$$\mathsf{CH}^1_{\mathrm{alg}}(C \times C')_{\mathbb{Q}} \subset \mathsf{CH}^1_{\equiv}(C \times C')_{\mathbb{Q}}$$

so that we may set

$$C^1_{\equiv,\mathrm{alg}}(C \times C')_{\mathbb{Q}} := \mathsf{CH}^1_{\equiv}(C \times C')_{\mathbb{Q}} / \mathsf{CH}^1_{\mathrm{alg}}(C \times C')_{\mathbb{Q}}$$

and then we get (see also the theorem of Weil[7] in Appendix A-2):

$$
\begin{aligned}
\mathrm{Hom}_{\mathrm{AV}}(J(C), J(C'))/\text{isog.} \ &\xrightarrow{\sim}\ \mathsf{CH}^1(C \times C')_{\mathbb{Q}} / \mathsf{CH}^1_{\equiv}(C \times C')_{\mathbb{Q}} \\
&=\ C^1_{\mathrm{alg}}(C \times C')_{\mathbb{Q}} / C^1_{\equiv,\mathrm{alg}}(C \times C')_{\mathbb{Q}}.
\end{aligned}
$$

---

[6]Recall the notation $\mathfrak{C}^{\natural}$ for the pseudo-abelian completion of a category $\mathfrak{C}$

[7]Since we use $\mathbb{Q}$–coefficients in our case, in applying the theorem of Weil we have to work up to isogeny.

Hence the theorem of Weil also holds, up to isogeny, if we replace $\mathsf{CH}^1(-)$ by $C^1_{\mathrm{alg}}(-)$. Now we can repeat the proof of (b) by replacing everywhere $\mathsf{CH}^1(-)$ by $C^1_{\mathrm{alg}}(-)$. Since by the theorems of Matsusaka and of Voisin-Voevodsky we have

$$
\begin{aligned}
\mathsf{CH}^1_{\mathrm{alg}}(C \times C')_{\mathbb{Q}} &= \mathsf{CH}^1_{\otimes}(C \times C)_{\mathbb{Q}} \\
&= \mathsf{CH}^1_{\mathrm{hom}}(C \times C')_{\mathbb{Q}} = \mathsf{CH}^1_{\mathrm{num}}(C \times C')_{\mathbb{Q}},
\end{aligned}
$$

we get the same in $\mathsf{Mot}_{\otimes}$, $\mathsf{Mot}_{\mathrm{hom}}$ and $\mathsf{Mot}_{\mathrm{num}}$.

2.8. Manin's Identity Principle

How do we detect that a correspondence f from X to Y is trivial? For homological equivalence this is easy: if $H(f) = 0$, then $f = 0$ (use the Künneth formula). However, if we work with rational equivalence it is not true that for $f \in \mathsf{Corr}_{\mathrm{rat}}(X \times Y) = \mathsf{CH}(X \times Y)$ we have $f = 0$ if the induced map $f_* : \mathsf{CH}(X) \to \mathsf{CH}(Y)$ is the zero map (even if we test this for all field extensions $K \supset k$). For instance, take $X = Y$ an elliptic curve E and $a, b, c, d \in E(k)$ four different points in 'general position'. Then $\{(a) - (b)\} \times \{(c) - (d)\} \in \mathsf{CH}^2(E \times E)$ is not zero, but it acts as zero on the Chow groups of E (and similarly on $\mathsf{CH}(E_K)$).

However, if we require that f is zero for all T-valued points, instead of ordinary points, then $f = 0$. Precisely, for $T \in \mathsf{SmProj}(k)$, put $X(T) := \mathsf{Corr}(T, X) = \mathsf{CH}(X \times T; \mathbb{Q})$ and for $f \in \mathsf{Corr}(X, Y)$ introduce the homomorphism

$$
\begin{aligned}
f_T : X(T) &\longrightarrow Y(T), \\
\alpha &\longmapsto f_T(\alpha) = f \circ \alpha.
\end{aligned}
$$

With this notation we have

THEOREM (Manin's identity principle [**Manin**, §3, p. 450], [**Ful**, ex. 16.1.12]). *Let $f, g \in \mathsf{Corr}(X, Y)$. Then the following are equivalent:*

    (1) *$f = g$;*
    (2) *$f_T = g_T$ for all $T \in \mathsf{SmProj}(k)$;*
    (3) *$f_X = g_X$.*

In particular,

f is an isomorphism $\iff f_T$ is an isomorphism for all $T \iff f_X$ is an isomorphism.

Proof: The implications 1) $\implies$ 2) $\implies$ 3) $\implies$ 1) are trivial (take $\alpha = \Delta(X)$ for the last step). $\qquad\square$

Remark. Although the proof is trivial, the principle is powerful indeed! The reason is that by Lieberman's lemma (Lemma 2.1.3) for $\alpha \in \mathsf{CH}(T \times X)$ we have

$$
f_T(\alpha) = f \circ \alpha = (\mathrm{id}_T \times f)_*(\alpha)
$$

so we have for $f, g \in \mathsf{Corr}(X, Y)$

$$
\mathsf{ch}(X) \xrightarrow{\mathsf{ch}(f) = \mathsf{ch}(g)} \mathsf{ch}(Y) \iff \mathsf{CH}(T \times X) \xrightarrow{(\mathrm{id}_T \times f)_* = (\mathrm{id}_T \times g)_*} \mathsf{CH}(T \times Y) \; \forall T. \quad (16)
$$

Hence $f = g$ as correspondences if and only if f and g act in the same way on Chow groups *universally*, i.e. after base change to the T-points and not merely on the "usual" points.

We illustrate the Manin principle with the following

EXAMPLES 2.8.1. (0) If $\varphi : X \to Y$ is a morphism of degree d, then $\Gamma_\varphi \circ {}^\mathsf{T}\Gamma_\varphi = d.\,\mathrm{id}_Y$. We have seen this already in Example 2.3 (vi), but it also follows from the identity principle [**Manin**, p. 450].

(1) (See [**Manin**, §7, p. 456], [**Scholl**, p. 171]) Let E be a locally free sheaf of rank $(m+1)$ on $S \in \mathsf{SmProj}(k)$ and let $\pi : \mathbb{P}_S(E) \to S$ be the associated projective bundle and ξ the tautological line bundle. Then there is an isomorphism of motives in $\mathsf{Mot}_{\mathrm{rat}}$

$$\mathsf{ch}(\mathbb{P}_S(E)) \xrightarrow{\sim} \bigoplus_{i=0}^{m} \mathsf{ch}(S)(-i).$$

This can be seen as follows. There is a well known isomorphism for Chow groups

$$\mathsf{CH}(\mathbb{P}_S(E)) \xrightarrow[\lambda]{\sim} \bigoplus_{i=0}^{m} \mathsf{CH}(S)[\xi^i]$$

and both λ and its inverse μ are given by correspondences. Moreover, this remains true universally, i.e. after an arbitrary base change $T \to \mathrm{Spec}\, k$. Therefore, since

$$(\mathrm{id}_T \times \mu) \circ (\mathrm{id}_T \times \lambda) = \mathrm{id} \quad \text{on } \mathsf{CH}(T \times \mathbb{P}_S(E)) \qquad \cdot$$

and similarly in the other direction, by (16) it follows that

$$\mathsf{ch}(\mathbb{P}_S(E)) \xrightarrow[\lambda]{\sim} \bigoplus_{i=0}^{m} \mathsf{ch}(S)(-i)$$

and μ are isomorphisms in $\mathsf{Mot}_{\mathrm{rat}}$ which are inverse to each other.

This gives the promised example of non-isomorphic varieties with the same motive: $S \times \mathbb{P}^m$ and $P_S(E)$ both give the same motive $\bigoplus_{i=0}^{m} \mathsf{ch}(S)(-i)$. Note that such projective bundles are mutually birationally equivalent.

(2) As an another application Manin [loc. cit., § 9, p. 461, resp. p. 172] shows that if $Y = \mathrm{Bl}_Z X$, the blow up of X along a smooth codimension $(m+1)$ submanifold Z, then

$$\mathsf{ch}(Y) = \mathsf{ch}(X) \oplus \bigoplus_{i=1}^{m} \mathsf{ch}(Z)(-i). \qquad (17)$$

CHAPTER 3

On Grothendieck's Standard Conjectures

In this chapter k is an algebraically closed field and X is a smooth, projective variety over k.

3.1. The Standard Conjectures

Up to now we have not used any conjectures; the construction of motives is very simple. However, motives also should enjoy a number of "hoped for" properties. In order to prove these, Grothendieck, around 1964 formulated a number of conjectures which he called **Standard Conjectures**:

- the Künneth conjecture $C(X)$;
- the conjectures of Lefschetz type $B(X)$ and $A(X, L)$;
- the conjecture of Hodge type $\mathrm{Hdg}(X)$
- conjecture $D(X)$, the equality of numerical and homological equivalence.

These conjectures are not independent of each other; for relations among them, see section 3.1.4.

For $k = \mathbb{C}$ these conjectures would follow from a special case of the Hodge conjecture.

The standard conjectures are discussed below, except the last one which has already been discussed in Chapter 1 (see 1.2.19). Together these imply the existence of a "universal" cohomology theory given by $\mathrm{Mot}_{\mathrm{num}}(k)$. In particular, the latter would have to be an abelian semi-simple category. That this is true is Jannsen's theorem [**Jann92**]. The standard conjectures would also imply the Weil conjectures. The latter have been proven by Deligne [**Del74b**], who avoided the standard conjectures.

To explain the standard conjectures, we fix a Weil cohomology theory $H(X)$ with F-coefficients, where F is a field of characteristic zero, e.g. $\mathbb{Q}$ or $\mathbb{Q}_\ell$. Recall (property (6) of a Weil cohomology theory) that we have the cycle class map

$$\mathsf{CH}^i(X)_{\mathbb{Q}} \xrightarrow{\gamma_X} \mathrm{Im}(\gamma_X) = \mathsf{A}^i(X) \subset H^{2i}(X). \tag{18}$$

Elements $z \in \mathsf{A}^i(X)$ are called *algebraic classes*.

3.1.1. The Künneth conjecture. Let $\Delta(X) \subset X \times X$ be the diagonal and consider its class

$$\gamma_{X \times X}(\Delta(X)) \in H^{2d}(X \times X) = \bigoplus_{i=0}^{2d} H^{2d-i}(X) \otimes H^i(X),$$

where the last equality is the Künneth decomposition. We write

$$\Delta_i^{\mathrm{topo}} \in H^{2d-i}(X) \otimes H^i(X)$$

35

for the i-th Künneth component.

Conjecture $C(X)$:   the Künneth components Δ_i^{topo} are algebraic, i.e. there is a cycle class $\Delta_i \in \mathsf{CH}^d(X \times X)_{\mathbb{Q}}$ with $\gamma_{X \times X}(\Delta_i) = \Delta_i^{\mathrm{topo}}$.

This conjecture is known (trivially) for varieties that admit an algebraic cell decomposition: projective spaces, Grassmannians and flag varieties; cf. [**Klei68**, Example 1.2.6]. It is trivial for curves. The conjecture is also known for surfaces and abelian varieties. (In these cases it is deduced from conjecture $B(X)$; see the discussion in section 3.1.2, in particular Remarks 3.1.1 (2).) For X defined over a finite field Katz and Messing [**Katz-Me**] proved it. In fact it is a consequence of Deligne's proof of the Weil conjectures.

Finally, note that if $k = \mathbb{C}$, the conjecture $C(X)$ follows from the Hodge conjecture for X.

3.1.2. Standard conjectures of Lefschetz type.

Let X_d be a smooth variety given with an explicit embedding into some projective space, and let Y be a smooth hyperplane section. Let $\gamma_X(Y) \in A^1(X) \subset H^2(X)$ be its class and let

$$L : H^i(X) \to H^{i+2}(X), \quad \alpha \mapsto \alpha \cup \gamma_X(Y)$$

be the *Lefschetz operator*. Its r-fold iteration is denoted L^r.

Recall that H being a Weil cohomology, we assume hard Lefschetz

$$L^{d-i} : H^i(X) \xrightarrow{\sim} H^{2d-i}(X), \quad 0 \le i \le d,$$

or, equivalently (putting $j = d - i$):

$$L^j : H^{d-j}(X) \xrightarrow{\sim} H^{d+j}(X), \quad 0 \le j \le d.$$

Note that L (and hence every L^j) is given by an algebraic cycle

$$L = \Delta(Y) \in \mathsf{CH}^{d+1}(X \times X).$$

Using the hard Lefschetz property one can define a unique linear map $\Lambda : H^i(X) \to H^{i-2}(X)$ $(2 \le i \le 2d)$ in cohomology which makes the following diagrams commute.

$i = d - j,\ 0 \le j \le d - 2$:

$$
\begin{array}{ccc}
H^{d-j}(X) & \xrightarrow[\sim]{L^j} & H^{d+j}(X) \\
\Big\downarrow{\scriptstyle \Lambda} & & \Big\downarrow{\scriptstyle L} \\
H^{d-j-2}(X) & \xrightarrow[\sim]{L^{j+2}} & H^{d+j+2}(X)
\end{array}
$$

$i = d + 1$:

$$
H^{d-1}(X) \;\underset{\Lambda}{\overset{L}{\rightleftarrows}}\; H^{d+1}(X)
$$

$i = d + j,\ 2 \le j \le d$:

$$
\begin{array}{ccc}
H^{d-j+2}(X) & \xrightarrow[\sim]{L^{j-2}} & H^{d+j-2}(X) \\
\Big\uparrow{\scriptstyle L} & & \Big\uparrow{\scriptstyle \Lambda} \\
H^{d-j}(X) & \xrightarrow[\sim]{L^j} & H^{d+j}(X),
\end{array}
$$

and of course, we can iterate $\Lambda^r = \underbrace{\Lambda \circ \cdots \circ \Lambda}_{r \text{ times}}$.

Remark. (1) Λ is almost an inverse of L. For instance, in the first diagram Λ is an inverse on the image of L inside $H^{d-j}(X)$.

(2) A *topological correspondence* is an element of $H^*(X \times X)$. The linear map $\Lambda : H^*(X) \to H^*(X)$ can be viewed as an element of $H^*(X)^\vee \otimes H^*(X) \simeq H^*(X) \otimes H^*(X) \subset H^*(X \times X)$ (by Poincaré-duality and the Künneth formula), and hence as a topological correspondence.

The standard conjecture of Lefschetz-type uses this last interpretation:

Conjecture $B(X)$: the topological correspondence Λ is algebraic, i.e. $\Lambda = \gamma_{X \times X}(Z)$ for some $Z \in \mathsf{CH}^{d-1}(X \times X)_{\mathbb{Q}}$.

If $B(X)$ holds, then clearly also the iterates Λ^r are algebraic. Note that for $r = d - 1$ the cycle class Λ^{d-1} is indeed algebraic and given by a divisor class [**Klei68**, Thm. 2A9.5].

Remarks 3.1.1. (1) For $k = \mathbb{C}$ the cycle Λ is Hodge, and hence $B(X)$ would follow from the Hodge conjecture. So this is an "old" conjecture, but Grothendieck emphasized its special role for motives.

(2) For general algebraically closed fields there are equivalent statements and several consequences. See [**Klei68, Klei94**]. we only mention:
- $B(X)$ is independent of the projective embedding used to define L and Λ;
- $B(X)$ implies $C(X)$.
- $B(X)$ implies the following conjecture $A(X, L)$. Consider the commutative diagram

$$\begin{array}{ccc} H^{2i}(X) & \xrightarrow[\sim]{L^{d-2i}} & H^{2d-2i}(X) \\ \uparrow & & \uparrow \\ \mathsf{A}^i(X) & \hookrightarrow & \mathsf{A}^{d-i}(X) \end{array}$$

where, as before (see (18)) $\mathsf{A}^i(X) = \mathrm{Im}(\gamma_X) \subset H^{2i}(X)$, i.e. those cohomology classes which are algebraic. By hard Lefschetz the lower map is injective.

Conjecture $A(X, L)$: the lower map in the above diagram is an isomorphism.

An equivalent formulation is the following.

Conjecture $A(X, L)'$: the cup product pairing

$$\mathsf{A}^i(X) \times \mathsf{A}^{d-i}(X) \to \mathbb{Q}$$

is non-degenerate.

Note that the $\mathsf{A}^i(X)$ are $\mathbb{Q}$-vector spaces and that due to the compatibility of cup-product and intersection the above pairing actually takes values in $\mathbb{Q}$.

Current status: the conjecture $B(X)$ is trivially true for projective space, for Grassmannians and for curves. For surfaces it is classical (Grothendieck, Kleiman; see [**Klei68**, Corollary 2A10]). For abelian varieties it is due to Lieberman and Kleiman [loc. cit., Theorem 2A11].

3.1.3. Standard conjecture of Hodge type. By the Hard Lefschetz theorem $L^{d-i}(X) : H^i(X) \xrightarrow{\sim} H^{2d-i}(X)$ if $i \leq d$, but in general $L^{d-i+1}(X)$ will have a non-trivial kernel:

$$P^i(X) := \mathrm{Ker}(L^{d-i+1} : H^i(X) \to H^{2d-i+2}(X)) \ (i\text{-th primitive cohomology})$$

and we may speak of the *primitive algebraic classes*

$$\mathsf{A}^i_{\mathrm{prim}}(X) := \mathsf{A}^i(X) \cap P^{2i}(X).$$

Cup-product induces for $i \leq d/2$ a pairing

$$
\begin{array}{ccc}
\mathsf{A}^i_{\mathrm{prim}}(X) \times \mathsf{A}^i_{\mathrm{prim}}(X) & \longrightarrow & \mathbb{Q} \\
(x, y) & \longmapsto & (-1)^i \, \mathrm{Tr} \circ (L^{d-2i}(x) \cup y)
\end{array}
\tag{19}
$$

and we have

Conjecture Hdg(X)(=standard conjecture of Hodge type): the pairing (19) is positive definite.

The current status is as follows:

(1) If $\mathrm{char}(k) = 0$ conjecture $\mathrm{Hdg}(X)$ is true by Hodge theory: by the *Lefschetz principle* one may assume that $k = \mathbb{C}$. Comparison of the cohomology theory with Betti-cohomology then shows the result in view of the Hodge-Riemann bilinear relations.

(2) In arbitrary characteristic it is known to hold for surfaces [**Segre**], [**Groth58**].

3.1.4. Consequences of and relations between the standard conjectures. For a full discussion, see [**Klei68**, **Klei94**]. We mention a couple of consequences having to do with conjecture $D(X)$ 1.2.19:

(1) $D(X) \implies A(X, L)$: by $D(X)$ we have $\mathsf{A}^i(X) = \mathsf{Z}^i(X)/\mathsf{Z}^i_{\mathrm{num}}(X)$ and the pairing

$$\mathsf{Z}^i(X)/\mathsf{Z}^i_{\mathrm{num}}(X) \times \mathsf{Z}^{d-i}(X)/\mathsf{Z}^{d-i}_{\mathrm{num}}(X) \to \mathbb{Q}$$

is non-degenerate by the definition of numerical equivalence.

(2) By [**Klei94**, Prop. 5.1], if $\mathrm{Hdg}(X)$ holds then $D(X) \iff A(X, L)$ hence in particular, $B(X)$ together with $\mathrm{Hdg}(X)$ imply $D(X)$. So in characteristic 0, since $\mathrm{Hdg}(X)$ holds, we have that $B(X)$ for all X is equivalent to $D(X)$ for all X.

(3) Furthermore [**Klei94**, Cor. 4.2.]:

$$B(X) \text{ for all } X \iff A(X, L) \text{ for all } X.$$

(4) $B(X)$ together with $\mathrm{Hdg}(X)$ imply that $\mathrm{Mot}_{\mathrm{num}}$ is an abelian semi-simple category. This consequence has been proved unconditionally in 1992 [**Jann92**]. We sketch his proof in § 3.2.

For the situation concerning divisors see § A-1 in Appendix A.

3.2. Jannsen's Theorem

We work over an algebraically closed field k and we let $\sim$ be any adequate equivalence relation. We fix also a field F containing $\mathbb{Q}$.

THEOREM 3.2.1 (Jannsen (1991)). *The following assertions are equivalent*

(1) $\mathsf{Mot} := \mathsf{Mot}_\sim$ *is an abelian semi-simple category;*
(2) $\sim$ *equals numerical equivalence;*
(3) *for all* $X_d \in \mathsf{SmProj}(k)$ *the* F*-algebra*

$$\mathsf{Corr}^0_\sim(X, X)_F = \mathsf{Corr}^0(X, X) \otimes_\mathbb{Q} F$$

is a finite-dimensional semi-simple F-algebra.

Remarks. 1) This is undoubtedly the most important progress on motives after Grothendieck, but we are still far away from the standard conjectures; in particular $D(X)$ is still not known.

2) An abelian category is semi-simple if every object is a finite direct sum of *simple* objects, i.e. objects without non-trivial subobjects.

Proof:

(1) $\implies$ (2) This goes by contradiction. Suppose that Mot is abelian and semi-simple but $\mathsf{Z}_\sim(-) \neq \mathsf{Z}_{\mathrm{num}}(-)$. As we mentioned before (subsection 1.2.1) $\mathsf{Z}_\sim(-) \subseteq \mathsf{Z}_{\mathrm{num}}(-)$ for every adequate equivalence relation. Suppose that there exists $Z \in \mathsf{Z}^i_{\mathrm{num}}(X)$ but $Z \notin \mathsf{Z}^i_\sim(X)$. Then, writing $\mathrm{pt} = \operatorname{Spec} k$, there is a morphism $f : \mathbf{1} = (\mathrm{pt}, \mathrm{id}, 0) \to h_\sim(X)(i) = (X, \mathrm{id}, i)$ in Mot given by $f = Z \in \mathsf{Corr}^i_\sim(\mathrm{pt}, X) = C^i_\sim(X)$ and since $Z \notin \mathsf{Z}^i_\sim(X)$ we have $f \neq 0$. Since Mot is abelian and semi-simple, there is a morphism $g : h_\sim(X)(i) \to \mathbf{1}$ such that $g \circ f = \mathrm{id}_{\mathrm{pt}}$. Such g is given by $W \in C^{d-i}_\sim(X)$. Then $g \circ f = (\mathrm{pr}_{13})_*(\mathrm{pr}^*_{12}(\mathrm{pt} \times Z) \cap \mathrm{pr}^*_{23}(W \times \mathrm{pt})) = \#(Z \cdot W) \cdot \mathrm{pt}$, where the intersection is on $\mathrm{pt} \times X \times \mathrm{pt} = X$ and $\#(Z \cdot W)$ is the intersection number on X. Since $g \circ f = \mathrm{id}_{\mathrm{pt}}$ we have $\#(Z \cdot W) = 1$. However, then Z is *not* numerically equivalent to zero, a contradiction. Hence $\mathsf{Z}_\sim(-) = \mathsf{Z}_{\mathrm{num}}(-)$.

(2) $\implies$ (3) This is the main step, i.e. if $\sim$ equals numerical equivalence, then $\mathsf{Corr}^0_\sim(X, X)_F$ is a finite dimensional and semi-simple algebra. Let X be a d-dimensional smooth projective variety. Recall (formula (18) in section 3.1) that we introduced the algebraic classes $\mathsf{A}^i(X) = \operatorname{Im}(\gamma^i_X) \subset H^{2i}(X)$ where $H(-)$ is some Weil cohomology with F-coefficients. For simplicity we take étale cohomology and $F = \mathbb{Q}_\ell$ with $\ell \neq p$, the characteristic of k. Put now

$$\mathsf{B}^i(X) := C^i_{\mathrm{num}}(X)_\mathbb{Q}.$$

We have a surjection $\mathsf{A}^i(X) \twoheadrightarrow \mathsf{B}^i(X)$. We know that $\dim_F \mathsf{A}^i(X)$ is finite, but a priori we cannot deduce from this anything about $\dim_\mathbb{Q} \mathsf{A}^i(X)$. Nevertheless we do have

LEMMA 3.2.2 ([**Klei68**, Thm. 3.5]).

$$\dim_\mathbb{Q} \mathsf{B}^i(X) \leq b_{2i}(X) = \dim H^{2i}_{\text{ét}}(X, \mathbb{Q}_\ell) < \infty.$$

Proof of Lemma 3.2.2: Choose $\alpha_1, \ldots, \alpha_m \in \mathsf{Z}^{d-i}(X)$ whose classes in $H^{2d-2i}_{\text{ét}}(X, \mathbb{Q}_\ell)$ form a maximal set of $\mathbb{Q}_\ell$-linearly independent elements in the image of the cycle class map $\gamma_X : \mathsf{Z}^{d-i}(X) \to H^{2d-2i}_{\text{ét}}(X, \mathbb{Q}_\ell)$. Clearly $m \leq b_{2d-2i}(X) = b_{2i}(X)$. For any 0-cycle γ we set $\#(\gamma := \deg(\gamma)$, Consider the linear map

$$\begin{aligned} \lambda \ : \ \mathsf{Z}^i(X) \ &\to \ \mathbb{Z}^m \\ \beta \ &\mapsto \ (\#(\beta \cdot \alpha_1), \ldots, \#(\beta \cdot \alpha_m)). \end{aligned}$$

We *claim* that $\operatorname{Ker}\lambda = \mathsf{Z}^i_{\mathrm{num}}(X)$. Clearly $\mathsf{Z}^i_{\mathrm{num}}(X) \subset \operatorname{Ker}\lambda$. Conversely, let $\alpha \in \mathsf{Z}^{d-i}(X)$ and write $\gamma_X(\alpha) = \sum \nu_j \gamma_X(\alpha_j)$ with $\nu_j \in \mathbb{Q}_\ell$. One has

$$\begin{aligned}
\#(\beta \cdot \alpha) &= \operatorname{Tr}\left(\gamma_X(\alpha) \cup \gamma_X(\beta)\right) = \\
&= \sum_j \nu_j \operatorname{Tr}\left(\gamma_X(\beta) \cup \gamma_X(\alpha_j)\right) = \\
&= \sum_j \nu_j \cdot \#(\beta \cdot \alpha_j)
\end{aligned}$$

by the compatibility of the cycle map with intersections and cup product. So if $\beta \in \operatorname{Ker}(\lambda)$ it follows that $\#(\beta \cdot \alpha) = 0$ for all $\alpha \in \mathsf{Z}^{d-i}(X)$, i.e. $\beta \in \mathsf{Z}^i_{\mathrm{num}}(X)$. hence λ induces a injection $\mathsf{B}^i(X) \hookrightarrow \mathbb{Q}^m$ which proves the claim. $\square$

LEMMA 3.2.3 (Key Lemma). *The $\mathbb{Q}$-algebra $\mathsf{B}^d(X \times X)$ is semi-simple.*

Proof of the Key Lemma: First a couple of remarks from the theory of non-commutative rings for which we refer to [**Bourb**, Livre 2 Chap 8] and [**Pie**]. If $F \supset \mathbb{Q}$ and R is a finite dimensional F-algebra, then R is artinian and the Jacobson radical $J(R)$ is the largest two-sided nilpotent ideal of R. It is the smallest two-sided ideal I making R/I semi-simple [**Bourb**, p.66]. So one must show that $J(\mathsf{B}^d(X \times X)) = 0$ and in fact, it suffices to prove this after base change to $\mathbb{Q}_\ell$ since $J(R \otimes_F F_1) = J(R) \otimes_F F_1$ for any overfield F_1 of F [loc.cit., p.70]. For simplicity, put $\mathsf{A} = \mathsf{A}^d(X \times X) \otimes_{\mathbb{Q}} \mathbb{Q}_\ell$, $\mathsf{B} = \mathsf{B}^d(X \times X) \otimes_{\mathbb{Q}} \mathbb{Q}_\ell$ and $J = J(\mathsf{B})$, $J' = J(\mathsf{A})$. Note that there is a surjection

$$\Phi : \mathsf{A} \twoheadrightarrow \mathsf{B}$$

and we need to show that $J = 0$. We *claim* that $\Phi(J') = J$. To see this, note that since Φ is surjective, $\Phi(J')$ is a two-sided nilpotent ideal in B and hence $\Phi(J') \subset J$ [loc.cit., p. 69]. On the other hand A/J' is a semi-simple $\mathbb{Q}_\ell$-algebra and hence so is its image $\mathsf{B}/\Phi(J')$ [**Pie**, Corollary p. 42], which is the middle algebra in the following sequence of surjections

$$\mathsf{A}/J' \twoheadrightarrow \mathsf{B}/\Phi(J') \twoheadrightarrow \mathsf{B}/J.$$

But J is the smallest ideal making the quotient B/J semi-simple. Hence $J \subset \Phi(J')$.

Next, since $J = J(C^d_{\mathrm{num}}(X \times X)) \otimes F$ we can start with $f \in J(C^d_{\mathrm{num}}(X \times X))$ which lifts to $f' \in J'$ and hence f' is nilpotent in A. Now take any $\tilde{g} \in \mathsf{Z}^d(X \times X)$ and let $g \in \mathsf{A}$ be its image. Now comes the crucial point: apply the Lefschetz trace formula[1] to f' and g:

$$\operatorname{Tr}\left(f' \cup {}^{\mathsf{T}}g\right) = \sum_{i=0}^{2d} (-1)^i \operatorname{Tr}_{H^i}(f' \circ g) \tag{20}$$

where ${}^{\mathsf{T}}g$ is the transpose and $f' \circ g$ operates on $H^i_{\text{ét}}(X, \mathbb{Q}_\ell)$. Since $f' \in J'$, also $f' \circ g \in J'$ since J' is a two-sided ideal. Hence $f' \circ g$ is nilpotent and in particular all traces are zero. So the right-hand side of (20) vanishes. But $\operatorname{Tr}\left(f' \cup {}^{\mathsf{T}}g\right) = \#(f \cdot {}^{\mathsf{T}}g)$ on $X \times X$ and $\tilde{g}$ is arbitrary in $\mathsf{Z}^d(X \times X)$. So $f = 0$ in $\mathsf{B}(X \times X)$ and hence $J = 0$. $\square$

$(3) \implies (1)$.
This is rather formal and we don't repeat the argument; it depends on the following two lemmas [**Jann92**]:

---

[1] Since $H(X)$ is a Weil cohomology theory, the Lefschetz trace formula is valid; cf. [**Klei68**, 1.3.6]

LEMMA 3.2.4. *Condition (3) implies that for every $M \in \mathsf{Mot}_\sim$ one has:*
(1) $\dim \mathrm{End}_{\mathsf{Mot}_\sim}(M) < \infty$;
(2) $\mathrm{End}_{\mathsf{Mot}_\sim}(M)$ *is a semi-simple F-algebra.*

LEMMA 3.2.5. *If $\mathfrak{A}$ is an F-linear pseudo-abelian category such that for every $M \in \mathfrak{A}$ its endomorphism algebra $\mathrm{End}_{\mathfrak{A}}(M)$ is a finite dimensional semi-simple F-algebra, then $\mathfrak{A}$ is an abelian semi-simple category.*

$\square$

Remarks. (1) The main ingredient of Jannsen's proof is the Lefschetz trace formula, which was known to Grothendieck.
(2) The crucial conjecture $D(X)$ remains open!
(3) So it follows from Jannsen's theorem that in general $\mathsf{Mot}_{\mathrm{rat}}$ is not an abelian semi-simple category. In fact it is not even abelian, see [**Scholl**, Cor. 3.5, p. 173].

CHAPTER 4

Finite Dimensionality of Motives

4.1. Introduction

Start with a smooth projective variety X defined over an algebraically closed field k. Recall that for divisors the Chow group $\mathsf{CH}^1(X) = \mathrm{Pic}(X)$ can be described as follows. The subgroup $\mathsf{CH}^1_{\mathrm{hom}}(X) = \mathsf{CH}^1_{\mathrm{alg}}(X)$ is an abelian variety, the *Picard variety* $\mathrm{Pic}^0_{\mathrm{red}}(X)(k)$ and the quotient $\mathsf{CH}^1(X)/\mathsf{CH}^1_{\mathrm{alg}}(X)$, the *Néron-Severi group* is a finitely generated group.

However, for cycles in higher codimension the picture is entirely different. Consider for instance the case of 0-cycles on a surface S. Recall from Chapter 1 Appendix A-3.1 that

$$\mathsf{CH}^2(S) \supset \mathsf{CH}^2_{\mathrm{hom}}(S) = \mathsf{CH}^2_{\mathrm{alg}}(S) = \mathsf{A}_0(S) \supset T(S),$$

where $\mathsf{A}_0(S)$ are the degree 0 zero-cycles and where

$$T(S) = \mathrm{Ker}\left[\mathrm{alb}_S : \mathsf{A}_0(S) \to \mathrm{Alb}(S)\right].$$

As mentioned in this appendix, the Albanese kernel $T(S)$ is infinite dimensional in some precise geometric sense if $H^2(S)_{\mathrm{trans}} \neq 0$. So it came as a surprise that Kimura [**Kimu**] showed that $\mathsf{CH}^2(S)$ still can be finite dimensional in a different motivic sense.

4.2. Preliminaries on Group Representations

Let G be a finite group. The *group ring* $R = \mathbb{Q}[G]$ consists of formal linear combinations $r = \sum_{g \in G} r(g)g$ with rational coefficients $r(g)$. If we replace these by complex coefficients we get $R_{\mathbb{C}}$. There are $h =$ (the number of conjugacy classes of G) non-isomorphic irreducible representations W_j with characters χ_j, $j = 1, \ldots, h$. Consider

$$e_j = \frac{\dim W_j}{\|G\|} \sum_{g \in G} \overline{\chi_j(g)} \cdot g.$$

Then in $R_{\mathbb{C}}$ we have (see e.g. [**Ful-Ha**, p. 17] or [**Serre77**, p. I-22]:

$$e_i \cdot e_j = \begin{cases} 0 & \text{if } i \neq j \\ e_i & \text{if } i = j. \end{cases}$$

$$\sum_i e_i = e_G = 1_R,$$

i.e. the e_i are idempotents decomposing 1 and they are orthogonal in $R_{\mathbb{C}}$.

Let us apply this to the symmetric group $\mathfrak{S}_n$ of permutations of n elements

$$\sigma = \begin{pmatrix} 1 & 2 & \cdots & n \\ \sigma(1) & \sigma(2) & \cdots & \sigma(n) \end{pmatrix}.$$

43

It is known (see e.g. [**Ful-Ha**, p. 46] or [**Cu-Re**, p. 47]) that the irreducible representations of $\mathfrak{S}_n$ correspond in a one to one manner to the *partitions* of n, i.e. to collections of positive integers

$$\lambda = (\lambda_1, \ldots, \lambda_s), \quad \lambda_1 \geq \cdots \geq \lambda_s$$

summing up to n. Let W_λ be the representation corresponding to λ. Then, using λ as indices in the preceding formulas we get:

$$e_\lambda \ = \ \frac{\dim W_\lambda}{\|G\|} \sum_{g \in G} \overline{\chi_\lambda(g)} \cdot g \tag{21}$$

$$e_\lambda \cdot e_\mu \ = \ \begin{cases} 0 & \text{if } \lambda \neq \mu \\ e_\lambda & \text{if } \lambda = \mu. \end{cases}$$

$$\sum_\mu e_\mu \ = \ 1_R$$

$$R_{\mathbb{C}} \ = \ \bigoplus_\lambda Re_\lambda, \quad Re_\lambda = \underbrace{W_\lambda \oplus \cdots \oplus W_\lambda}_{\dim W_\lambda \text{ times}}$$

Particular cases are:

(1) $\lambda = (n)$ corresponds to the trivial representation $\sigma(v) = v$ and

$$e_{\text{sym}} := e_{(n)} = \frac{1}{n!} \sum_{\sigma \in \mathfrak{S}_n} \sigma.$$

(2) $\lambda = (1, \ldots, 1)$ corresponds to the alternating representation, i.e. $\sigma(v) = \text{sgn}\sigma \cdot v$ and

$$e_{\text{alt}} := e_{(1,\ldots,1)} = \frac{1}{n!} \sum_{\sigma \in \mathfrak{S}_n} \text{sgn}(\sigma) \cdot \sigma.$$

4.3. Action of the Symmetric Group on Products

The action of the symmetric group $\mathfrak{S}_n$ on $X^n = \underbrace{X \times \cdots \times X}_{n \text{ times}}$ is as follows:

$$\begin{array}{ccc} X^n & \to & X^n \\ (x_1, \ldots, x_n) & \mapsto & (x_{\sigma(1)}, \ldots, x_{\sigma(n)}). \end{array}$$

Let $\Gamma_\sigma(X)$ be the graph of this map, i.e., the subvariety of $X^n \times X^n$ consisting of points of the form $(x_1, \ldots, x_n, x_{\sigma(1)}, \ldots, x_{\sigma(n)})$.

More generally, for any $r = \sum_\sigma r(\sigma)\sigma \in R$ we get a $\mathbb{Q}$-correspondence $\Gamma_r(X) \in \text{Corr}_{\widetilde{}}^0(X^n)$ given by

$$\Gamma_r(X) = \sum_{\sigma \in \mathfrak{S}_n} r(\sigma)[\Gamma_\sigma(X)].$$

Note that the product in R corresponds to composition of correspondences:

$$\Gamma_{rs}(X) = \Gamma_r(X) \circ \Gamma_s(X).$$

Hence

$$\boxed{d_\lambda(X) := \Gamma_{e_\lambda}(X) : X^n \to X^n} \tag{22}$$

is a projector.

The same story holds for motives $M = (X, p, m) \in \mathsf{Mot}_\sim$ and their products

$$M^{\otimes n} = (\underbrace{X \times \cdots \times X}_{n \text{ times}}, \underbrace{p \times \cdots \times p}_{n \text{ times}}, nm).$$

The group $\mathfrak{S}_n$ operates through the first factor $\underbrace{X \times \cdots \times X}_{n \text{ times}}$ and if we abbreviate

$\underbrace{p \times \cdots \times p}_{n \text{ times}} = p^{\otimes n}$, a small calculation then shows

$$\Gamma_\sigma(M) := \Gamma_\sigma(X) \circ p^{\otimes n} = p^{\otimes n} \circ \Gamma_\sigma(X)$$

so that $\Gamma_\sigma(M)$ is a morphism in $\mathsf{Mot}_\sim$.

As for varieties, more generally any $r = \sum r(\sigma)\sigma \in R$ operates on $M^{\otimes n}$ and one has

$$\boxed{\Gamma_r(M) := \sum_\sigma r(\sigma)\Gamma_\sigma(M) \in \mathrm{Hom}_{\mathsf{Mot}_\sim}(M^{\otimes n}, M^{\otimes n}).}$$

Let us now look at the projectors $d_\lambda(X) = \Gamma_{e_\lambda}(X)$:

LEMMA 4.3.1. *Let $M = (X, p, m) \in \mathsf{Mot}_\sim$. Then*

(1) $d_\lambda(X) \circ p^{\otimes n} = p^{\otimes n} \circ d_\lambda(X)$ *is a projector for X^n;*
(2) *the $d_\lambda(X) \circ p^{\otimes n}$ decompose $p^{\otimes n}$;*
(3) $d_\lambda(X) \circ p^{\otimes n} \perp d_\mu(X) \circ p^{\otimes n}$ *if $\lambda \neq \mu$.*

Proof: This follows formally from $\Gamma_r(M) \circ \Gamma_s(M) = \Gamma_{rs}(M)$ and the fact that $p^{\otimes n}$ commutes with the Γ_r. $\qquad\square$

DEFINITION 4.3.2. Let $M = (X, p, m) \in \mathsf{Mot}_\sim$ and λ a partition of n. Put

$$\mathbb{T}_\lambda M := (X^n, d_\lambda \circ p^{\otimes n}, nm) \in \mathsf{Mot}_\sim. \tag{23}$$

In particular

$$\begin{aligned}
\mathrm{Sym}^n(M) &:= \mathbb{T}_{(n)}M = (X^n, d_{\mathrm{sym}} \circ p^{\otimes n}, nm) \\
{\textstyle\bigwedge}^n(M) &:= \mathbb{T}_{(1,\ldots,n)}M = (X^n, d_{\mathrm{alt}} \circ p^{\otimes n}, nm)
\end{aligned}$$

It follows immediately from the definitions that

$$\mathsf{CH}(\mathbb{T}_\lambda M) = \mathrm{Im}(d_\lambda) \subset \mathsf{CH}(M^{\otimes n})$$

and similarly for the cohomology groups. In particular therefore

$$\begin{aligned}
\mathsf{CH}(\mathrm{Sym}^n(M)) &= \mathrm{Im}(d_{\mathrm{sym}}) \subset \mathsf{CH}(M^{\otimes n}) \\
\mathsf{CH}({\textstyle\bigwedge}^n M) &= \mathrm{Im}(d_{\mathrm{alt}}) \subset \mathsf{CH}(M^{\otimes n}).
\end{aligned}$$

Let H be a Weil cohomology theory. Recall from Definition 1.2.13 that H satisfies the super-commutativity rule

$$b \cup a = (-1)^{ij} a \cup b, \quad \text{if } a \in H^i(X), \ b \in H^j(X).$$

We shall write

$$\boxed{H^{\mathrm{even}} = H^+(X) = \oplus H^{2i}(X), \quad H^{\mathrm{odd}}(M) = H^-(X) = \oplus H^{2i+1}(X).}$$

PROPOSITION 4.3.3. *Let $M = (X, p, 0) \in \mathsf{Mot}_\sim$. Then*

$$\begin{aligned}
H(\mathrm{Sym}^n(M)) &= \bigoplus_{i+j=n} \mathrm{Sym}^i H^+(M) \otimes {\textstyle\bigwedge}^j H^-(M) \\
H({\textstyle\bigwedge}^n M) &= \bigoplus_{i+j=n} {\textstyle\bigwedge}^i H^+(M) \otimes \mathrm{Sym}^j H^-(M).
\end{aligned}$$

Proof: This follows from the Künneth formula and the super-commutativity rule. For a proof of the first formula in the case $M = \mathrm{ch}(X)$ see [**BanRol**, Prop. 3.2.3] (cf. also [**Kimu**, p. 181-182]). $\qquad\square$

It implies the following basic result:

COROLLARY 4.3.4.   (1) *If $H^+(M) = 0$ then $H(\mathrm{Sym}^n(M)) = \Lambda^n H^-(M)$ and hence $H(\mathrm{Sym}^n(M)) = 0$ if $n > \dim H(M)$.*
  (2) *If $H^-(M) = 0$ then $H(\Lambda^n(M)) = \Lambda^n H^+(M)$ and hence $H(\Lambda^n(M)) = 0$ if $n > \dim H(M)$.*

4.4. Dimension of Motives

Inspired by Corr. 4.3.4, for $M = (X, p, m) \in \mathsf{Mot}_\sim$ Kimura and independently O'Sullivan have introduced:

DEFINITION 4.4.1.   (1) *M is evenly finite dimensional if $\Lambda^n M = 0$ for some $n > 0$, i.e. $d_{\mathrm{alt}}(X) \circ p^{\otimes n} \sim 0$ and $\dim M$ then is defined as the maximal number n for which $\Lambda^n(M) \neq 0$;*
  (2) *M is oddly finite dimensional if $\mathrm{Sym}^n(M) = 0$ for some $n > 0$ i.e. $d_{\mathrm{sym}}(X) \circ p^{\otimes n} \sim 0$ and $\dim(M)$ is defined as the maximal number n for which $\mathrm{Sym}^n(M) \neq 0$;*
  (3) *M is finite dimensional if $M = M_+ \oplus M_-$ with M_+ evenly finite dimensional and M_- oddly finite dimensional. In this case[1] $\dim(M) = \dim(M_+) + \dim(M_-)$.*

Remarks. (1) Note that (X, p, m) is finite dimensional if and only if $(X, p, 0)$ is finite dimensional, and that both motives have the same dimension.
(2) We have $\mathfrak{S}_n \subset \mathfrak{S}_{n+1}$ and $e_{\mathrm{sym}}^{(n+1)} = r \cdot e_{\mathrm{sym}}^{(n)}$ for some suitable $r \in R(\mathfrak{S}_{n+1})$, and similarly for e_{alt}. For example, let $n = 3$, $\tau = (12)$, $\sigma = (123)$, then

$$
\begin{aligned}
e_{\mathrm{sym}}^{(3)} &= (1 + \sigma + \sigma^2)(1 + \tau) = (1 + \sigma + \sigma^2)e_{\mathrm{sym}}^{(2)} \\
e_{\mathrm{alt}}^{(3)} &= (1 + \sigma + \sigma^2)(1 - \tau) = (1 + \sigma + \sigma^2)e_{\mathrm{alt}}^{(2)}.
\end{aligned}
$$

It follows that

$$
\begin{aligned}
p^{\otimes(n+1)} \circ d_{\mathrm{sym}}^{(n+1)}(X) &= p^{\otimes(n+1)} \circ \Gamma_r \circ (\Gamma_{d_{\mathrm{sym}}^n}(X) \times \mathrm{id}) \\
&= \Gamma_r \circ (p^{(n)} \times p) \circ (\Gamma_{d_{\mathrm{sym}}^n}(X) \times \mathrm{id})
\end{aligned}
$$

and hence $\mathrm{Sym}^n(M) = 0$ implies $\mathrm{Sym}^{n+1}(M) = 0$ and similarly for $\Lambda^n M$.

EXAMPLES 4.4.2. $\mathbf{1} \in \mathsf{Mot}_\sim$ is evenly finite dimensional and its dimension is 1 because $\Lambda^2 \mathbf{1} = (\mathrm{Spec}\, k \times \mathrm{Spec}\, k, \mathrm{id} - \mathrm{id}, 0) = 0$. By part (1) of the above remark, the Lefschetz motive $\mathbb{L} = (\mathbb{P}^1, \mathbb{P}^1 \times e, 0) \simeq (\mathrm{Spec}\, k, \mathrm{id}, -1)$ is also evenly finite dimensional and $\dim \mathbb{L} = 1$. More generally, $(X_d, p_0, 0)$ and $(X_d, p_{2d}, 0)$ are evenly finite dimensional and have dimension 1 for the same reason.

---

[1] A priori this depends on the decomposition, but we shall see later Prop. 5.3.3 that this is well-defined

4.5. The Sign Conjecture and Finite Dimensionality

We let $\sim\;=$ homological equivalence with respect to our chosen Weil cohomology. We have

CONJECTURE 4.5.1. [**Sign conjecture** $S(X)$] The sum of all the even projectors $(\Delta_{2i}^{\text{topo}})_X$ as well as the sum of all the odd projectors $(\Delta_{2i+1}^{\text{topo}})_X$ are algebraic, i.e., there exist algebraic cycles Δ^+ and Δ^- such that

$$\gamma_{X\times X}(\Delta^+) \;=\; \sum_i (\Delta_{2i}^{\text{topo}})_X$$

$$\gamma_{X\times X}(\Delta^-) \;=\; \sum_i (\Delta_{2i+1}^{\text{topo}})_X.$$

Remark 4.5.2. The Künneth conjecture $C(X)$ (see § 3.1.1) implies the sign conjecture.

LEMMA 4.5.3. *If $S(X)$ holds then $h_{\text{hom}}(X)$ has finite dimension (and also $h_\sim(X)$ whenever $\mathsf{Z}_{\text{hom}}(X) \subset \mathsf{Z}_\sim(X)$).*

Proof: Since the condition $\mathsf{Z}_{\text{hom}}(X) \subset \mathsf{Z}_\sim(X)$ implies the existence of a functor $\mathsf{Mot}_{\text{hom}} \to \mathsf{Mot}_\sim$ it suffices to prove this for homological equivalence. Write

$$(X, \text{id}, 0) = \underbrace{(X, \Delta^+, 0)}_{M_+} \oplus \underbrace{(X, \Delta^-, 0)}_{M_-}.$$

Also let $\dim H(M_+) = \dim H^+(X) = n$. Then $\dim H(\bigwedge^{n+1} M_+) = 0$. We claim that the cycle class of $d_{\text{alt}}^{(n+1)}(X)\circ\Delta_+^{\otimes(n+1)}$ is 0, i.e. vanishes in $\text{Hom}_{\mathsf{Mot}_{\text{hom}}}(M_+, M_-)$. This claim follows from the fact that the cohomology functor

$$H : \mathsf{Mot}_{\text{hom}}^{\text{eff}} \to \text{Gr Vect}_F$$

is faithful. To see this, let $M = (X, p, 0)$ and $N = (Y, q, 0)$. Consider $f \in \text{Hom}_{\mathsf{Mot}_{\text{hom}}}(M, N)$ If $H(f) = 0$ then f operates trivially in cohomology and using the Künneth decomposition (formula (5) in § 1.2.D) this implies that $f \sim_{\text{hom}} 0$, i.e. $f = 0$ as a morphism in $\mathsf{Mot}_{\text{hom}}^{\text{eff}}$.

A similar argument applies to M_-. $\qquad\square$

4.6. Curves

Now we shall work with Chow motives, i.e. in $\mathsf{Mot}_{\text{rat}}$. Recall (§ 2.7) that for any curve C we have a decomposition

$$
\begin{array}{ccccccc}
\mathsf{ch}(C) & = & \mathsf{ch}^0(C) & \oplus & \mathsf{ch}^1(C) & \oplus & \mathsf{ch}^2(C) \\
& & \| & & \| & & \| \\
& & (C, p_0 = e \times C) & & (C, p_1 = \Delta - p_0 - p_2) & & (C, p_2 = C \times e)
\end{array}
$$

Of course $\mathsf{ch}^+(C) = \mathsf{ch}^0(C)\oplus\mathsf{ch}^2(C)$ is evenly finite dimensional and has dimension 2. That $\mathsf{ch}^-(C) = \mathsf{ch}^1(C)$ is oddly finite dimensional is also true:

THEOREM 4.6.1 ([**Serm, Kü93, Kimu**]). *The motive $\mathsf{ch}^1(C)$ is oddly finite dimensional of dimension $2g$. Hence $\mathsf{ch}(C)$ itself is also finite dimensional.*

Proof: First of all, $H(\text{Sym}^{2g}(\mathsf{ch}^1(C))) = \bigwedge^{2g} H^1(C) \neq 0$ so the dimension must be $\geq 2g$. We show that $\text{Sym}^{2g+1}(\mathsf{ch}^1(C)) = 0$.

To do this we introduce some notation. Recall than $p_1^{\otimes n}$ is a cycle on $C^n \times C^n$ and we then put

$$\alpha_n := d_{\mathrm{sym}}^{(n)}(C) \circ p_1^{\otimes n} = \left(\frac{1}{n!} \sum_{\sigma \in \mathfrak{S}_n} \Gamma_\sigma\right) \circ p_1^{\otimes n} \in \mathsf{CH}^n(C^n \times C^n).$$

Observe that $\mathrm{Sym}^n(\mathsf{ch}^1(C)) = (C^n, \alpha_n, 0)$ and hence the assertion we claim, namely $\mathrm{Sym}^{2g+1}(\mathsf{ch}^1(C)) = 0$ just means that $\alpha_{2g+1} = 0$.

To show the latter, we shall reduce it to vanishing of a suitable element on the $(2g+1)$-fold symmetric product of C (see Step I below). This leads us to introduce the n-fold symmetric product $S^n C := C^n / \mathfrak{S}_n$ which is a smooth variety (since $\dim C = 1$) with the natural projection

$$\varphi_n : C^n \to S^n C.$$

Then α_n induces the correspondence

$$\begin{aligned}
\beta_n &= \frac{1}{n!} \cdot (\varphi_n)_* \circ \alpha_n \circ \varphi_n^* \\
&= \frac{1}{n!} (\varphi_n \times \varphi_n)_* \alpha_n \in \mathsf{CH}^n(S^n C \times S^n C),
\end{aligned}$$

where we used the Lieberman identity (Lemma 2.1.3) for the last equality.
Step I. $\alpha_n = 0 \iff \beta_n = 0$ (and hence it suffices to prove that $\beta_{2g+1} = 0$).
This assertion follows from the following Lemma:

> LEMMA 4.6.2. *For all positive integers n we have*
> (1) *β_n is a projector of $S^n C$;*
> (2) *$\alpha_n = \frac{1}{n!} \varphi_n^* \circ \beta_n \circ (\varphi_n)_*$;*
> (3) *$\mathrm{Sym}^n(\mathsf{ch}^1(C)) \simeq (S^n C, \beta_n)$ where the morphisms $\frac{1}{n!}(\varphi_n)_*$ and φ_n^* induce isomorphisms of motives which are inverse to each other.*

Proof of the Lemma: (1) and (2) are straightforward calculations using $\varphi_n^* \circ (\varphi_n)_* = n! \left(\frac{1}{n!} \sum_{\sigma \in \mathfrak{S}_n} \Gamma_\sigma\right) = n! d_{\mathrm{sym}}^{(n)}(C)$ plus the fact that α_n is a projector.
(3) goes as follows. We use § 2.2.2. Note that $\mathrm{Sym}^n(\mathsf{ch}^1(C)) = (C^n, \alpha_n)$ and so we must check whether the two morphisms of *motives* $\frac{1}{n!}\beta_n \circ \varphi_n^* \circ \alpha_n$ and $\alpha_n \circ \varphi_n^* \circ \beta_n$ are inverse isomorphisms. So we consider consider the two compositions

$$\alpha_n \circ \varphi_n^* \circ \beta_n \circ \underbrace{\frac{1}{n!}(\varphi_n)_* \circ \alpha_n}_{\alpha_n} : C^n \to C^n$$

and

$$\beta_n \circ \underbrace{\frac{1}{n!}(\varphi_n)_* \circ \alpha_n \circ \varphi_n^*}_{\beta_n} \circ \beta_n : S^n C \to S^n C.$$

Note that, β_n and α_n being projectors, the compositions as indicated evaluate to α_n, respectively β_n, the identity morphisms of the respective motives.
Step II. The description of the Chow group of $S^n C$ if $n > 2g - 2$.
Under the assumption $n > 2g - 2$, as is well known, the Abel-Jacobi map $\pi : S^n C \to J(C)$ is a projective fibre bundle with fibres of dimension $m = n - g$. Hence, if we let $\xi_n := \mathcal{O}(1)$ be the tautological bundle for the projective fibre bundle one has (see e.g. [**Ful**]):

$$\mathsf{CH}(S^n C)) \simeq \mathsf{CH}(J(C))[1, \xi_n, \ldots, \xi_n^m].$$

More precisely, any $\beta \in \mathsf{CH}^r(S^nC)$ can be written

$$\beta = \pi^*(a_r) + \pi^*(a_{r-1})\xi_n + \cdots + \pi^*(a_{r-m})\xi_n^m, \quad \text{with } \alpha_j \in \mathsf{CH}^j(\mathrm{J}(C)).$$

This relation make it possible to show the vanishing of β using the projection formula for π and the fact that for dimension reasons $\pi_*\xi_n^\ell = 0$ when $\ell < m$ and $= 1$ when $\ell = m$. Indeed, suppose that by induction we have already shown that $a_{r-k} = 0$ for $k = \ell + 1, \ldots, m$. Then it follows that $\pi_*(\beta \cdot \xi_n^\ell) = 0$ implies $a_{r-\ell} = 0$ as well. In other words, β vanishes if the direct images $\pi_*(\beta \cdot \xi_n^\ell)$ vanish for $\ell = 0, \ldots, m$. So this becomes a calculation on $\mathrm{J}(C)$. Below we need a slightly more involved variant of this idea.

Step III. Completion of the proof.

Consider now the two projections

$$\mathrm{pr}_1, \mathrm{pr}_2 : S^{2g+1}C \times S^{2g+1}C \to S^{2g+1}C.$$

Then, because of Step II (with $r = 2g + 1, m = g + 1$), we can write

$$\beta_{2g+1} = \sum_{i,j=0}^{g+1}(\pi{\circ}\mathrm{pr}_1 \times \pi{\circ}\mathrm{pr}_2)^* a_{ij} \cdot \mathrm{pr}_1^*\xi^{g+1-i} \cdot \mathrm{pr}_2^*\xi^{g+1-j}, \quad \xi := \xi_{2g+1}$$
$$a_{ij} \in \mathsf{CH}^{i+j-1}(\mathrm{J}(C) \times \mathrm{J}(C))$$

As explained in Step II, the proof that $\beta_{2g+1} = 0$ can reduced to showing that for $0 \le i, j \le g + 1$ one has:

$$a_{ij} = (\pi{\circ}\mathrm{pr}_1 \times \pi{\circ}\mathrm{pr}_2)_* \left(\mathrm{pr}_1^*\xi^i \cdot \mathrm{pr}_2^*\xi^j \cdot \beta_{2g+1}\right) = 0. \tag{24}$$

This is clear for $i = j = 0$ since $a_{0,0} \in \mathsf{CH}^{-1}(\mathrm{J}(C) \times \mathrm{J}(C)) = 0$.[2] For the other i, j we have either $i > 0$ or $j > 0$ (or both). Let us assume $i > 0$. Now apply the projection formula for $\pi \times \pi$. Then it suffices to show that

$$\mathrm{pr}_1^*\xi \cdot \beta_{2g+1} = 0 \quad \text{on } S^{2g+1}C \times S^{2g+1}C.$$

For the remainder of the proof we set

$$\varphi = \varphi_{2g+1} : C^{2g+1} \to S^{2g+1}C$$

and recall that $\beta_{2g+1} = \frac{1}{(2g+1)!}(\varphi \times \varphi)_*\alpha_{2g+1}$ so that

$$\mathrm{pr}_1^*\xi \cdot \beta_{2g+1} = \mathrm{pr}_1^*\xi \cdot \frac{1}{(2g+1)!}(\varphi \times \varphi)_*\alpha_{2g+1}. \tag{25}$$

We now invoke a classical result. Fix $e \in C$ and consider for all $n > 0$ the inclusion

$$S^n(C) \xrightarrow{i} S^{n+1}C$$
$$(x_1, \ldots, x_n) \mapsto (x_1, \ldots, x_n, e).$$

Then, if $n > 2g - 2$, by [**Schwar**] one has $\xi_{n+1} = i_*[S^nC]$. In particular, one has

$$\mathrm{pr}_1^*\xi = S^{2g}C \times e \times S^{2g+1}.$$

Applying the projection formula for $\varphi \times \varphi$ we thus get

$$\mathrm{pr}_1^*\xi \cdot \beta_{2g+1} = (\varphi \times \varphi)_*(\varphi \times \varphi)^* \left[S^{2g}C \times e \times S^{2g+1}\right] \cdot \alpha_{2g+1}$$

so that (25) becomes

$$\mathrm{pr}_1^*\xi \cdot \beta_{2g+1} = \frac{1}{(2g+1)!}(\varphi \times \varphi)_* \left[\left(\sum_j (C \times \cdots \times \underset{\substack{\uparrow \\ j\text{-th factor}}}{e} \times \cdots \times C) \times C^{2g+1}\right)\right] \cdot \alpha_{2g+1}$$

---

[2] We used that β_{2g+1} is a cycle of dimension $(2g + 1)$, so this argument definitely does not work for β_n with $n < 2g + 1$

Since α_{2g+1} comes from p_1, by symmetry the term in the last line which is written in the last parentheses is zero if $[e \times C^{2g} \times C^{2g+1}] \cdot \sigma(p_1 \times \cdots \times p_1) = 0$ for all $\sigma \in \mathfrak{S}_n$ which is the case since

$$(e \times C) \cdot p_1 = (e \times C) \cdot (\Delta(C) - e \times C - C \times e) = 0. \qquad \square$$

CHAPTER 5

Properties of Finite Dimensional Motives

For this chapter the unpublished notes [**Jann03**] of a course of U. Jannsen in Tokyo have been very helpful.

5.1. Sums and Tensor Products

PROPOSITION 5.1.1. 1) *Let M and N be evenly (oddly) finite dimensional motives. Then $M \oplus N$ is evenly (oddly) finite dimensional. Conversely, if $M \oplus N$ is evenly (oddly) finite dimensional then M and N are evenly (oddly) dimensional.*

2) If M and N are finite dimensional, then $M \oplus N$ is finite dimensional; moreover $\dim(M \oplus N) \leq \dim M + \dim N$.

Proof: We only look at the case of evenly finite dimensional motives. The equality

$$\bigwedge^n (M \oplus N) = \bigoplus_{r+s=n} \bigwedge^r M \otimes \bigwedge^s N$$

shows that if $\bigwedge^{\ell+1} M = 0 = \bigwedge^{k+1} N$, then $\bigwedge^{k+\ell+1}(M \oplus N) = 0$; conversely, if $\bigwedge^n(M \oplus N) = 0$, then all the terms on the right hand side vanish and so, in particular, $\bigwedge^n M = \bigwedge^n N = 0$. $\qquad\square$

Remark. Conversely we shall see (Cor. 5.4.6) that if $M \oplus N$ finite dimensional, then both M and N are finite dimensional.

Tensor products are much more involved; one needs some more representation theory of the symmetric group in addition to what we used in § 4.2. Below we recall the results that we shall use; these can be found in [**Ful-Ha**] or [**Cu-Re**].

$$\text{(i)} \quad \left\{ \begin{array}{c} \text{irred. repr. } W_\lambda \\ \text{of } \mathfrak{S}_n \end{array} \right\} \underset{1-1}{\Longleftrightarrow} \left\{ \begin{array}{c} \text{partitions } \lambda \\ \text{of } n \end{array} \right\} \underset{1-1}{\Longleftrightarrow} \left\{ \begin{array}{c} \text{Young diagrams} \\ \text{of weight } n. \end{array} \right\}$$

Here a partition $\lambda = (\lambda_1, \ldots, \lambda_s)$ corresponds to a diagram with s rows of lengths $\lambda_1, \ldots \lambda_s$. For instance $\lambda = (4, 2, 1)$ corresponds to the diagram

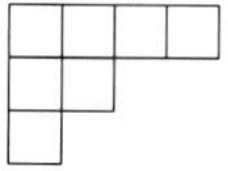

(ii) A *Young tableau* is obtained from a Young diagram by putting the integers $1, \ldots, n$ in the table. If you put them in this order starting on the left in the first row, one gets the *standard tableau*. Here are two examples; the first is the standard tableau.

1	2	3	4
5	6		
7			

7	1	3	6
2	4		
5			

To a Young tableau T one associates two subgroups of the symmetric group

$$R_\lambda(T) \quad := \quad \{\sigma \in \mathfrak{S}_n \mid \sigma \text{ only permutes elements in each row}\}$$
$$C_\lambda(T) \quad := \quad \{\sigma \in \mathfrak{S}_n \mid \sigma \text{ only permutes elements in each column}\}$$

and three associated elements in the group ring

$$a_\lambda(T) \quad := \quad \sum_{\sigma \in R_\lambda(T)} \sigma$$

$$b_\lambda(T) \quad := \quad \sum_{\sigma \in C_\lambda(T)} \mathrm{sgn}(\sigma)\sigma$$

$$c_\lambda(T) \quad := \quad a_\lambda(T)b_\lambda(T) \quad \text{the } \textit{Young symmetrizer} \text{ of } T.$$

EXAMPLES 5.1.2. 1) $\lambda = (n)$. Then $R_\lambda(T) = \mathfrak{S}_n$, $C_\lambda(T) = (\mathrm{id})$ and

$$c_\lambda = a_\lambda = \sum_{\sigma \in \mathfrak{S}_n} \sigma = e_{\mathrm{sym}}.$$

2) $\lambda = (1, \ldots, 1)$. Then $R_\lambda(T) = \mathrm{id}$, $C_\lambda(T) = \mathfrak{S}_n$ and

$$c_\lambda = b_\lambda = \sum_{\sigma \in \mathfrak{S}_n} \mathrm{sgn}(\sigma)\sigma = e_{\mathrm{alt}}.$$

Remark. Let V be a vector space. The symmetric group acts on $V^{\otimes n}$ by permuting the factors. One can show that, regardless of the numbers filled in the tableau, we have

$$\mathrm{Im}(a_\lambda) \simeq \mathrm{Sym}^{\lambda_1}(V) \otimes \cdots \otimes \mathrm{Sym}^{\lambda_s}(V)$$

and that

$$\mathrm{Im}(b_\lambda)) \simeq \textstyle\bigwedge^{\mu_1} V \otimes \cdots \otimes \bigwedge^{\mu_s} V$$

where μ is the partition dual to λ that one gets upon changing rows and columns of the associated Young diagram; cf. [**Ful-Ha**, p. 46]. For instance is dual

to 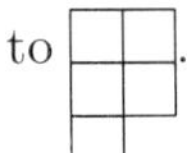 .

Facts from group theory. See [**Ful-Ha**, p. 46] or [**Cu-Re**, Thm. 28.15, p. 197]:

(1) $c_\lambda(T) \circ c_\lambda(T) = n_\lambda(T)c_\lambda(T)$, $n_\lambda(T) \neq 0$, in the group ring $R = \mathbb{Q}[\mathfrak{S}_n]$, i..e $c_\lambda(T)$ is almost an idempotent;

(2) $Rc_\lambda(T)$ is a minimal left ideal in R, hence it is an irreducible R-module;

(3) $Rc_\lambda(T) = Rc_\mu(T) \iff \lambda = \mu$, i.e. the ideal $Rc_\lambda(T)$ only depends on the partition and not on the tableau;

(4)
$$\sum_{T \in \lambda} Rc_\lambda(T) = \bigoplus W_\lambda^{\oplus \dim W_\lambda} \subset R,$$

i.e. W_λ, the irreducible representation corresponding to λ occurs precisely $\dim W_\lambda$ times;

(5) each of the orthogonal idempotents e_λ (see (21) in Chapter 4) is a linear combination of monomials in $c_\lambda(T)$.

Next, let us explain the crucial vanishing lemma. First, recall (cf. (22) in Chapter 4) the following notation. If $M = (X, p, m) \in \mathsf{Mot}_{\sim}(k)$ and λ is a partition of n we introduced (22) the correspondence $d_\lambda(M) := \Gamma_{e_\lambda}(M)$, the correspondence associated to e_λ and the motive (23):

$$\mathbb{T}_\lambda M := (X^n, d_\lambda(M) \circ p^{\otimes n}, nm).$$

LEMMA 5.1.3 (Vanishing Lemma [**Kimu**, Thm. 5.9]). *Let* $q \geq n$ *and* $\lambda = (\lambda_1, \ldots, \lambda_s)$ *a partition of* q. *Then*

(1) *if* $\mathrm{Sym}^{n+1}(M) = 0$ *and* $\lambda_1 > n$, *then* $\mathbb{T}_\lambda M = 0$;
(2) *if* $\bigwedge^{n+1} M = 0$ *and* $\lambda_{n+1} \neq 0$, *then* $\mathbb{T}_\lambda M = 0$.

Proof: We only prove (2); the argument for (1) being similar. Rearranging the numbering we can assume that

$$\text{dual of } T =$$

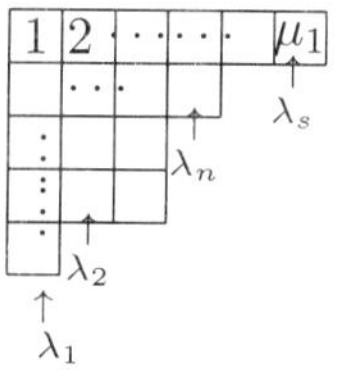

In other words: the tableau associated to the dual μ of the partition λ is the standard tableau. In the above tableau the duality aspect has been illustrated: the columns have length $\lambda_1, \ldots, \lambda_s$ with $s > n$.

Now $b_{\lambda, T} : M^{\otimes q} \to M^{\otimes q}$ and we have

$$\mathrm{Im}(b_{\lambda, T}) = \bigwedge{}^{\mu_1} M \otimes \cdots \otimes \bigwedge{}^{\mu_s} M.$$

Also $\lambda_{n+1} \neq 0$ hence $\mu_1 > n$ and since $\bigwedge^{n+1} M = 0$ we have that $b_\lambda(M^{\otimes q}) = 0$ so certainly $c_\lambda(M^{\otimes q}) = 0$. Since $e_\lambda = r \cdot c_\lambda$ for a certain $r \in R$, we get $d_\lambda(M^{\otimes q}) = \mathbb{T}_\lambda M = 0$. $\square$

THEOREM 5.1.4. *If* M *and* N *are finite dimensional, then* $M \otimes N$ *is finite dimensional. More precisely:*

(1) *If* M, N *are both evenly or oddly finite dimensional, then* $M \otimes N$ *is evenly finite dimensional;*
(2) *If* M *and* N *are finite dimensional of different parity, then* $M \otimes N$ *is oddly finite dimensional.*

Moreover, $\dim(M \otimes N) \leq \dim M \cdot \dim N$.

Proof: By way of example we shall prove that if $\mathrm{Sym}^{m+1}(M) = \mathrm{Sym}^{n+1}(N) = 0$, then $\bigwedge^{nm+1}(M \otimes N) = 0$. Write $q = mn + 1$. Then

$$(M \otimes N)^{\otimes q} \simeq M^{\otimes q} \otimes N^{\otimes q}$$
$$\bigoplus_{\lambda, \nu} d_\lambda M^{\otimes q} \otimes d_\nu N^{\otimes q},$$

where λ and ν run independently through all partitions of q. We need to prove $d_{\mathrm{alt}}(M \otimes N)^{\otimes q} = 0$, i.e.

$$d_{(1, \cdots, 1)}\left[\bigoplus_{\lambda, \nu} d_\lambda M^{\otimes q} \otimes d_\nu N^{\otimes q} = 0 \right]. \tag{26}$$

Since by [**Ful-Ha**, p. 61] one has

$$e_{(1,\dots,1)} \cdot e_\lambda \otimes e_\nu = \begin{cases} 0 & \text{unless } \nu = \mu, \text{ the dual of } \lambda \\ e_{(1,\cdots,1)} & \text{if } \nu = \mu \end{cases}$$

to prove (26) it suffices to see that $d_\lambda M^{\otimes q} \otimes d_\mu N^{\otimes q} = 0$. Now, if $\lambda_1 \geq m + 1$, then $d_\lambda M^{\otimes q} = 0$ by the vanishing lemma (1). If $\lambda_1 \leq m$ then $\lambda_{n+1} > 0$ and hence $\mu_1 \geq n + 1$ and then $d_\mu N^{\otimes q} = 0$, again by the vanishing lemma (1). $\square$

COROLLARY 5.1.5. *If $h_\sim(X)$ and $h_\sim(Y)$ are finite dimensional, then so is $h_\sim(X \times Y)$. In particular, if C is a curve then $\mathsf{ch}(C^n)$ is finite dimensional for all n.*

5.2. Smash Nilpotence of Morphisms

This section is a digression and a preparation for § 5.3, but it is of substantial interest itself.

DEFINITION 5.2.1. A morphism $f : M \to N$ in $\mathsf{Mot}_\sim(k)$ is called *smash-nilpotent* or *tensor-nilpotent* if for some integer $n > 0$ the cycle class

$$[\Gamma_f \times \cdots \times \Gamma_f] \in C_\sim(X \times Y)^n_{\mathbb{Q}} \simeq C_\sim(X^n \times Y^n)_{\mathbb{Q}}$$

vanishes. In other words, $f \sim_\otimes 0$ if for some $n > 0$ the class of the n-th exterior product $f^{\otimes n}$ in $C_\sim(X^n \times Y^n)_{\mathbb{Q}}$ vanishes. If $\sim = \sim_{\mathrm{rat}}$, this means that the associated correspondence $\Gamma_f \in \mathsf{Corr}_{\mathrm{rat}}(X \times Y)_{\mathbb{Q}}$ is smash nilpotent in the sense of Chapter 1.

Remarks 5.2.2. 1) If π_i is the projection from $X_1 \times \cdots \times X_n \times Y_1 \times \cdots \times Y_n$ onto the factor $X_i \times Y_i$ the exterior product of the n morphisms $f_i : X_i \to Y_i$, $i = 1, \dots, n$ can be explicitly given as

$$f_1 \otimes \cdots \otimes f_n = \pi_1^*(f_1) \cdot \dots \cdot \pi_n^*(f_n) \tag{27}$$

2) The notion of smash-nilpotence makes sense in any $\mathbb{Q}$–linear tensor category.

LEMMA 5.2.3. *Let $f, g : M \to N$ be smash-nilpotent, then $f + g$ and $f - g$ are smash-nilpotent.*

Proof: The proof of Proposition 1.2.10 goes through with $\sim_{\mathrm{rat}}$ replaced by $\sim$. $\square$
The main result in this section is:

THEOREM 5.2.4. *Let $f : M \to M$ in $\mathsf{Mot}_\sim(k)$ be smash nilpotent of order n, i.e. $f^{\otimes n} = 0$. Then $f^n = \underbrace{f \circ \cdots \circ f}_{n \text{ times}} = 0$. In other words: smash nilpotence implies nilpotence.*

This, in turn, is implied by the following

PROPOSITION 5.2.5. *Let $f : M \to N$ in $\mathsf{Mot}_\sim(k)$ be smash nilpotent of order n and let $g_i : N \to M$, $i = 1, \dots, n - 1$ be morphisms in $\mathsf{Mot}_\sim(k)$. Then the composition $f \circ g_{n-1} \circ f \circ \cdots \circ f \circ g_1 \circ f$ vanishes.*

Sketch of proof. Let $M = (X, p, -)$, $N = (Y, g, -)$, and pull everything back to the product $(X \times Y \times X) \times \cdots \times (X \times Y \times X)$ in the obvious way. This gives the result. For simplicity we illustrate the method by proving the following

LEMMA 5.2.6. *Suppose we have varieties X_1, X_2, X_3 and X_4 and correspondences $\Gamma_1 \in \mathsf{Corr}_\sim(X_1, X_2)_\mathbb{Q}$, $\Omega \in \mathsf{Corr}_\sim(X_2, X_3)_\mathbb{Q}$ and $\Gamma_2 \in \mathsf{Corr}_\sim(X_3, X_4)_\mathbb{Q}$. If $\Gamma_1 \times \Gamma_2 = 0$, then the composition $\Gamma_2 \circ \Omega \circ \Gamma_1$ vanishes.*

Proof: We denote by $\pi_{134} : X_1 \times X_2 \times X_3 \times X_4 \to X_1 \times X_3 \times X_4$ the projection and likewise for $\pi_{14} : X_1 \times X_3 \times X_4 \to X_1 \times X_4$ and $\pi_{13} : X_1 \times X_2 \times X_3 \to X_1 \times X_3$. Start out with the cycles

$$
\begin{aligned}
\alpha &= \{(\Gamma_1 \times X_3) \cdot (X_1 \times \Omega)\} \times X_4 \in C_\sim(X_1 \times X_2 \times X_3 \times X_4)_\mathbb{Q} \\
\beta &= X_1 \times \Gamma_2 \in C_\sim(X_1 \times X_3 \times X_4)_\mathbb{Q}.
\end{aligned}
$$

Now $\pi_{134} = \pi_{13} \times \mathrm{id}_{X_4}$ and

$$
\begin{aligned}
\alpha \cdot \pi_{134}^*(\beta) &= (\Gamma_1 \times X_3 \times X_4) \cdot (X_1 \times \Omega \times X_4) \cdot (X_1 \times X_2 \times \Gamma_2) \\
&= (\Gamma_1 \times \Gamma_2) \cdot (X_1 \times \Omega \times X_4) \\
&= 0.
\end{aligned}
$$

Apply the projection formula:

$$
0 = (\pi_{134})_*(\alpha \cdot \pi_{134}^*(\beta)) = (\pi_{134})_*(\alpha) \cdot \beta
$$

and note that

$$
(\pi_{134})_*(\alpha) = (\pi_{13})_* \{(\Gamma_1 \times X_3) \cdot (X_1 \times \Omega)\} \times \mathrm{id}_{X_4}(X_4) = (\Omega \circ \Gamma_1) \times X_4
$$

and hence on $X_1 \times X_3 \times X_4$ we have

$$
\{(\Omega \circ \Gamma_1) \times X_4\} \cdot (X_1 \times \Gamma_2) = 0
$$

Finally, apply the projection formula for π_{14} to this cycle; on $X_1 \times X_4$ we then get

$$
0 = (\pi_{14})_* \{(\Omega \circ \Gamma_1 \times X_4) \cdot (X_1 \times \Gamma_2)\} = \Gamma_2 \circ \Omega \circ \Gamma_1. \qquad \square
$$

5.3. Morphisms Between Finite Dimensional Motives

PROPOSITION 5.3.1. *Let M, N be two finite dimensional motives of different parity and let $f : M \to N$ be a morphism between them. Then f is smash nilpotent. More precisely, $f^{\otimes \ell} = 0$ if $\ell > \dim M \cdot \dim N$.*

Proof: Let λ, ν be two partitions of ℓ and consider

$$
M^{\otimes \ell} \xrightarrow{d_\lambda} M^{\otimes \ell} \xrightarrow{f^{\otimes \ell}} N^{\otimes \ell} \xrightarrow{d_\nu} N^{\otimes \ell}. \tag{28}
$$

Since the d_μ commute with correspondences the above map equals $f^{\otimes \ell} \circ d_\lambda \circ d_\nu$ which is automatically zero if $\lambda \neq \nu$. But since the d_μ are idempotents by construction, we get $f^{\otimes \ell} \circ d_\lambda$ if $\lambda = \nu$. Therefore it suffices to show that the map (28) vanishes for $\ell = nm + 1$ also for $\lambda = \nu$, where $n = \dim N$, $m = \dim M$. For simplicity, consider the case where $\bigwedge^{m+1} M = 0$ and $\mathrm{Sym}^{n+1}(N) = 0$. For the partition λ we have $\lambda_{m+1} > 0$ or $\lambda_{m+1} = 0$. In the first case, by the vanishing lemma 5.1.3 (2) we have $\mathbb{T}_\lambda M = 0$. In the second case $\lambda_1 > n$ and by the vanishing lemma (1), one has $\mathbb{T}_\lambda N = 0$. $\qquad \square$

COROLLARY 5.3.2. *Suppose that $M = (X, p, m)$ is evenly and oddly finite dimensional. Then $M = 0$.*

Proof: It suffices to show this in the case $M = (X, p, 0)$. Apply the previous Proposition to p (it is the identity morphism of M). It follows that p must be smash nilpotent and hence nilpotent. But then $p = 0$ since it is a projector, and so the motive vanishes. $\qquad \square$

We shall now give the main application of Prop. 5.3.1.

PROPOSITION 5.3.3. *Let $M = (X, p, m)$ be a finite dimensional motive. Then its decomposition $M = M_+ \oplus M_-$ into the odd and even parts is essentially unique, i.e. for any other decomposition $M = M'_+ \oplus M'_-$ into even and oddly finite dimensional motives, one must have $M_+ \simeq M'_+$ and $M_- \simeq M'_-$. In particular, $\dim M$ is well defined.*

Indication of the proof: Write

$$p = p_+ + p_- = p'_+ + p'_-.$$

Note that $p \circ p_\pm = p_\pm = p_\pm \circ$ and similarly for $p'_\pm$. Now consider

$$(p - p'_+) \circ p_+ : M_+ \to M'_-.$$

By Prop. 5.3.1 this is smash nilpotent, but it goes from X to X and hence it is nilpotent, say $\left[(p - p'_+) \circ p_+\right]^n = 0$. Now expanding, taking care of the non-commutativity, we get

$$p_+ = p_+^n = \underbrace{\mathrm{polynomial}(p_+, p'_+)}_{\lambda} \underbrace{p'_+ \circ p_+}_{\mu}.$$

Since $\mu : M_+ \to M'_+$ and $p_+ = \lambda \circ \mu$ we must have $\lambda : M'_+ \to M_+$. Since $\mu \circ \lambda = p_+ = \mathrm{id}_{M_+}$, by Chap. 2.§ 2.3 (viii) we then have

$$M'_+ = M_+ \oplus K,$$

where $K = \mathrm{Ker}(\lambda)$ is also evenly finite dimensional. For evenly finite dimensional motives the dimension is well-defined. Hence $\dim M'_+ = \dim M + \dim K$, i.e.

$$\dim M'_+ \geq \dim M_+.$$

But one can prove the opposite inequality in a similarly way so that we have in fact equality and $K = 0$ so that $M_+ \simeq M'_+$. $\square$

5.4. Surjective Morphisms and Finite Dimensionality

Let us work with Chow motives for simplicity.

DEFINITION 5.4.1 ([**Kimu**, Def. 6.5]). Let $f : M \to N$ be a morphism of motives. Then f is called *surjective* if for all smooth projective varieties Z the induced map

$$\mathsf{CH}(M \otimes \mathsf{ch}(Z))_{\mathbb{Q}} \to \mathsf{CH}(N \otimes \mathsf{ch}(Z))_{\mathbb{Q}}$$

is surjective.

EXAMPLE 5.4.2. Let $\phi : X_d \to Y_e$ be a dominant morphism and let $f = \mathsf{ch}(\phi) : \mathsf{ch}(X) \to \mathsf{ch}(Y)(e - d)$ be the morphism between motives induced by the closure in $X \times Y$ of the graph of ϕ. Then f is surjective. This applies in particular if f is a birational dominant *morphism*. Replacing ϕ by $\phi \times \mathrm{id}_Z$ we can forget about Z and it suffices to show that f is surjective

To see this, assume first that ϕ is a generically finite morphism of degree r (in particular $d = e$). Then we have morphisms of motives ϕ_* and ϕ^* for which $\phi_* \circ \phi^* = r\, \mathrm{id}$ and hence ϕ_* is onto.

The general case goes as follows. Construct a rational map $s : Y \to X$ which is generically a multi-section of $\phi : X \to Y$. The closure $X' \subset X$ of the image of Y under s thus lies generically finitely over Y, say $\phi' = \phi|X' \to Y$. For any irreducible subvariety $W \subset Y$ we have $\phi_*(X' \cdot \phi^*(W)) = rW$.

Dominant rational maps are in general *not* surjective. Consider for example the inverse $\phi : X \dashrightarrow Y = \mathrm{Bl}_P X$ of the blow up of a smooth projective variety X in a point P. Then ϕ is not surjective because $\mathsf{CH}^1(Y) = \mathsf{CH}^1(X) \oplus \mathbb{Z}E$ where E is the exceptional variety – and $E \not\subset \mathrm{Im}\,\phi_*$.

LEMMA 5.4.3 ([**Kimu**, Lemma 6.8]). *Let $f : M = (X, p, m) \to N = (Y, q, n)$ be a morphism of motives. The following conditions are equivalent:*

(1) *f is surjective;*
(2) *there exists a right inverse $g : N \to M$, i.e. $f{\circ}g = \mathrm{id}_N$.*
(3) *$q = f{\circ}s$ for some $s \in \mathsf{Corr}^0(Y, X)$.*

Proof: (2) $\implies$ (1): consider

$$\mathsf{CH}(N \otimes \mathsf{ch}(Z)) \xrightarrow{(g \times \mathrm{id}_Z)_*} \mathsf{CH}(M \otimes \mathsf{ch}(Z)) \xrightarrow{(f \times \mathrm{id}_Z)_*} \mathsf{CH}(N \otimes \mathsf{ch}(Z)).$$

Since $f{\circ}g = \mathrm{id}_N$ we get $(f \times \mathrm{id}_Z)_* \circ (g \times \mathrm{id}_Z)_* = \mathrm{id}_{M \otimes Z}$ on the Chow groups and hence $(f \otimes \mathrm{id}_Z)_*$ is a surjection.

(1) $\implies$ (2): take $Z = Y$ and ${}^\mathsf{T}q \in \mathsf{CH}(Y \times Y)$. By Lieberman's Lemma 2.1.3 we have $(q \times \mathrm{id}_Y)_*\, \mathrm{id}_Y = {}^\mathsf{T}q$ and so ${}^\mathsf{T}q \in \mathsf{CH}(N \times Y)$. So by assumption (1), there is some $r \in \mathsf{Corr}^0(X, Y)$ for which

$$r{\circ}\,{}^\mathsf{T}f = (f \times \mathrm{id}_Y)_* r = {}^\mathsf{T}q,$$

where the first equality follows again from Lieberman's Lemma. Now take $g := p{\circ}\,{}^\mathsf{T}r{\circ}q$. Clearly $p{\circ}g = g$, and $g{\circ}q = q$ so that indeed $g \in \mathrm{Hom}(N, M)$, but also

$$\mathrm{id}_N = q = q^2 = \underbrace{f{\circ}\,{}^\mathsf{T}r}_{q}{\circ}q = f{\circ}\underbrace{p{\circ}\,{}^\mathsf{T}r{\circ}q}_{g}$$

(2) $\implies$ (3) Take $s = g$.
(3) $\implies$ (2): Take $g = p{\circ}s{\circ}q$. $\square$

THEOREM 5.4.4 ([**Kimu**, Prop. 6.9]). *Let $f : M \to N$ be a surjective morphism of motives. If M is finite dimensional, then so is N.*

Proof: The easy case is when M is evenly finite dimensional or oddly finite dimensional. In these cases it follows immediately from the above Lemma since by § 2.3. (viii) $M = N \oplus K$ by surjectivity and M being evenly/oddly finite dimensional implies the same for N and K.

The general case goes as follows: Write $M = M_+ \oplus M_-$. One only needs to show that $N = N'_+ \oplus N'_-$ where $M_+ \twoheadrightarrow N'_+$ and $M_- \twoheadrightarrow N'_-$.

Write $M = (X, p, 0)$ and $N = (Y, q, 0)$ and consider f as a correspondence from X to Y with $f{\circ}p = f = q{\circ}f$. By Lemma 5.4.3 there exists $s \in \mathsf{Corr}^0(Y, X)$ for which $f{\circ}s = q$. Hence, since $p = p_+ + p_-$ we get

$$\begin{aligned}
q = q{\circ}q &= f{\circ}s{\circ}q \\
&= f{\circ}p{\circ}s{\circ}q \\
&= \underbrace{f{\circ}p_+{\circ}s{\circ}q}_{q'_+} + \underbrace{f{\circ}p_-{\circ}s{\circ}q}_{q'_-}
\end{aligned}$$

which gives two correspondences

$$q'_\pm = f{\circ}p_\pm{\circ}s{\circ}q : Y \to Y. \tag{29}$$

We first claim that these are endomorphisms of the *motive* N, i.e. that $q'_\pm \circ q = q'_\pm = q \circ q'_\pm$. While we clearly have $q'_\pm \circ q = q'_\pm$, the second equality requires verification:

$$q \circ q'_\pm = q \circ f \circ p_\pm \circ s \circ q = f \circ p \circ p_\pm \circ s \circ q = f \circ p_\pm \circ s \circ q = q'_\pm,$$

While it is not clear that the $q'_\pm$ are projectors, we shall show that certain polynomial expressions in $q'_\pm$ are indeed projectors. See the claims (1)–(4) below.

As a first step we show that the composition $q'_+ \circ q'_-$ is nilpotent: we have

$$q'_+ \circ q'_- = f \circ p_+ \circ s \circ q \circ f \circ p_- \circ s \circ q.$$

By Prop. 5.3.1 $p_+ \circ s \circ q \circ f \circ p_- : M_- \to M_+$ is smash nilpotent and so, by Prop. 5.2.5, the self-correspondence $f \circ p_+ \circ s \circ q \circ f \circ p_- \circ s \circ q$ of Y is nilpotent, say of order k. So

$$0 = (q'_+ \circ q'_-)^k = (q'_+ \circ (q - q'_+))^k = (q'_+ - q'^2_+)^k = q'^k_+ \circ q'^k_- \tag{30}$$

We put

$$q_+ := (q - (q'_-)^k)^k, \quad q_- := q - q_+.$$

Observe that $q \circ q_\pm = q_\pm = q_\pm \circ q$. We claim that the $q_\pm$ are the desired projectors. We first make the following remark: expanding $(q - t)^k$ where $q^2 = q$, an idempotent as before, and t a self-correspondence of Y with $q \circ t = t = t \circ q$ yields

$$(q - t)^k = q - P(t) \circ t \implies q - (q - t)^k = P(t) \circ t \tag{31}$$

for some (universal) polynomial P (which, given q only depends on k). We claim:

(1) $q_\pm = q'^k_\pm \circ r_\pm$ with $r_+ := P^k(q'_+)$, $r_- := P(q'^k_-)$ self-correspondences of Y;

(2) $q_+ \circ q'^k_+ = q'^k_+$;

(3) $q_\pm$ is an idempotent;

(4) $q_+ \circ q_- = 0$, $q_- \circ q_+ = 0$.

To prove these claims, apply (31) with $t = q'_+$ which yields $q_+ = (q - (q - q'_+)^k)^k = P^k(q'_+) \circ q'^k_+$, which proves assertion (1) for q_+. Applying it with $t = q'^k_-$ shows (1) for q_-:

$$q_- = q - q_+ = q - (q - q'^k_-)^k = P(q'^k_-) \circ q'^k_-.$$

Next, again using (31) together with the nilpotency relation (30) yields

$$q_+ \circ q'^k_+ = (q - P(q'^k_-) \circ q'^k_-) \circ q'^k_+ = q'^k_+,$$

which proves assertion (2) for q_+. It now follows from (1) and (2) that q_+ is an idempotent:

$$q_+ \circ q_+ = q_+ \circ q'^k_+ \circ r_+ = q'^k_+ \circ r_+ = q_+.$$

Since we have $q \circ q'_+ = q'_+ \circ q = q'_+$ the definition of q_+ implies that similar relations hold for q_+, i.e. $q \circ q_+ = q_+ \circ q = q_+$. Since q_+ is an idempotent, it then follows that $q_- = q - q_+$ is an idempotent as well. So (3) holds. Now (4) is immediate.

Finally, we show that $M_\pm$ surjects to $(Y, q_\pm)$ under f. Indeed, by (29) we have

$$q'_\pm = f \circ r'_\pm \quad \text{with } r'_\pm = p_\pm \circ s \circ q \implies q'^k_\pm = f \circ t_\pm, \quad t_\pm \in \mathrm{Corr}^0(Y, X)$$

and hence by assertion (1) we have likewise $q_\pm = f \circ s_\pm$, $s_\pm \in \mathrm{Corr}^0(Y, X)$. This, by Lemma 5.4.3, implies surjectivity under f. $\qquad\square$

This theorem has important consequences:

COROLLARY 5.4.5. *Let $f : X \to Y$ be a dominant morphism and suppose that* $\mathsf{ch}(X)$ *is finite dimensional. Then* $\mathsf{ch}(Y)$ *is finite dimensional.*

Proof: Using Example 5.4.2 this follows directly from Theorem 5.4.4. $\quad\square$

COROLLARY 5.4.6. *If $M \oplus N$ is finite dimensional, then so are M and N.*

Proof: The projections $M \oplus N \to M$ and $M \oplus N \to N$ are surjective by Lemma 5.4.3. Then apply Theorem 5.4.4. $\quad\square$

Since the Chow motive of a curve is finite dimensional (see § 4.6) this implies:

COROLLARY 5.4.7. *The Chow motive of a variety morphically dominated by a product of curves is finite dimensional. In particular this is true for an abelian variety.*

The last assertion follows since every abelian variety is dominated by the Jacobian of a curve C, say of genus g which in turn is dominated by a self-product of g times C.

An abstract consequence of the theorem is:

COROLLARY 5.4.8. *Every motive which is a direct summand of some tensor product of motives of curves is finite dimensional. These motives form a full tensor-subcategory inside the category of Chow motives.*

5.5. Finite Dimensionality and Nilpotence

THEOREM 5.5.1. *Let $M = (X, p, m)$ and let $f : M \to M$ be a morphism of Chow motives. Assume that M is either evenly finite-dimensional ($\bigwedge^n M = 0$) or oddly finite-dimensional ($\mathrm{Sym}^n(M) = 0$). We have*

   (a) *There exists a nonzero polynomial $G(T) \in \mathbb{Q}[T]$ of degree $n - 1$ with $G(f) = 0$.*

   (b) *If f is numerically trivial, then f is nilpotent (more precisely, $f^{n-1} = 0$).*

Remark. S. Kimura proved this in [**Kimu**] with $n - 1$ replaced by n. The stronger form is due to U. Jannsen; see [**Jann07**, Theorem 6.4.3].

The proof uses two auxiliary results. The first results from a repeated application of the projection formula to self correspondences of X, as in the proof of Lemma 5.2.6:

LEMMA 5.5.2. *Let $f_i \in \mathsf{Corr}(X, X)$, $i = 1, \ldots, n - 1$. Then*

$$f_{n-1} \circ \cdots \circ f_1 = (p_{1n})_* \left[p_{12}^* f_1 \cdot \ldots \cdot p_{n-1,n}^* f_{n-1} \right],$$

where the f_i are now considered as morphisms of Chow motives and where the p_{ij} are the projections

$$p_{ij} \;:\; X^n \to X \times X$$
$$(x_1, \ldots, x_n) \mapsto (x_i, x_j).$$

The second result reads as follows; the proof is left to the reader:

LEMMA 5.5.3. *Let V, W, X, Y be varieties, $f : V \to X$, $g : W \to Y$ two morphisms and $\alpha \in \mathsf{CH}(V)$, $\beta \in \mathsf{CH}(W)$. Consider the following diagram with the*

obvious projections

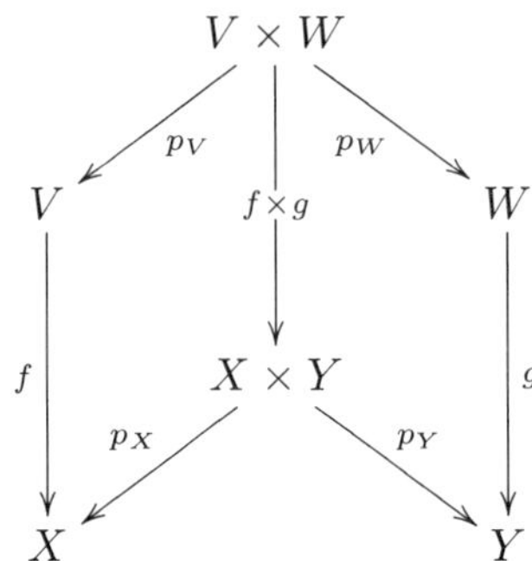

Then

$$(f \times g)_* \left(p_V^*(\alpha) \cdot p_W^*(\beta) \right) = p_X^* f_*(\alpha) \cdot p_Y^* g_*(\beta).$$

We now turn to the *proof of Theorem 5.5.1:* By applying a Tate twist, we may assume that $M = (X, p, 0)$. We give the proof in the case M is evenly finite-dimensional, i.e., $\bigwedge^n M = 0$. This means that

$$d_{\mathrm{alt}} \circ (p \times \ldots \times p) = \sum_{\sigma \in \mathfrak{S}_n} \mathrm{sgn}(\sigma) \Gamma_\sigma(X) \circ (p \times \ldots \times p) = 0.$$

Let $p_{ij} : X^{2n} \to X \times X$ be the projection to the i-th and j-th factor. As

$$p \times \ldots \times p = p_{1,n+1}^*(p) \cdot p_{2,n+2}^*(p) \cdot \ldots \cdot p_{n,2n}^*(p)$$

we obtain

$$\sum_{\sigma \in \mathfrak{S}_n} \mathrm{sgn}(\sigma) p_{1,n+\sigma(1)}^*(p) \cdot \ldots \cdot p_{n,n+\sigma(n)}^*(p) = 0. \tag{32}$$

Let f be an endomorphism of the motive M, i.e., $f \in \mathsf{Corr}^0(X, X)$ and $f \circ p = p \circ f = f$. Following Jannsen [loc.cit.], we consider the expression

$$\left. \sum_{\sigma \in \mathfrak{S}_n} \mathrm{sgn}(\sigma)(p_{1,n+1})_* (p_{1,n+\sigma(1)}^*(p) \cdot \ldots \cdot p_{n,n+\sigma(n)}^*(p) \cdot \\ \cdot p_{n+2,2}^*(f) \cdot \ldots \cdot p_{2n,n}^*(f)) \right\} \tag{33}$$

which vanishes by (32). We shall rewrite this expression as a polynomial in f of degree $n-1$ using Lemmas 5.5.2 and 5.5.3. Let us look at the summand corresponding to a permutation $\sigma \in \mathfrak{S}_n$. Write $\sigma = \sigma_1 \circ \sigma'$ in which σ_1 is a cycle in which 1 occurs and σ' is the product of the remaining cycles free from 1. Lets s be the length of the orbit of σ_1, i.e.

$$\mathrm{orbit}(\sigma_1) = \left\{ 1, \sigma(1), \ldots, \sigma^{s-1}(1) \right\}.$$

Rearranging the factors of (33) we find that the summand corresponding to σ is of the form

$$\mathrm{sgn}(\sigma)(p_{1,n+1})_*(\alpha \cdot \beta)$$

where

$$\alpha = p_{1,n+\sigma(1)}^*(p) \cdot p_{n+\sigma(1),\sigma(1)}^*(f) \cdot p_{\sigma(1),n+\sigma^2(1)}^*(p) \cdot p_{n+\sigma^2(1),\sigma^2(1)}^*(f) \cdot \ldots$$

$$\ldots \cdot p_{n+\sigma^{s-2}(1),\sigma^{s-1}(1)}^*(p) \cdot p_{n+\sigma^{s-1}(1),\sigma^{s-1}(1)}^*(f) \cdot p_{\sigma^{s-1}(1),n+1}^*(p)$$

and β is the intersection product of the remaining factors $p_{i,n+\sigma(i)}^*(p)$ or $p_{n+i,i}^*(f)$ with $i \notin \mathrm{orbit}_1(\sigma)$.

Write $X^{2n} = V \times W$ with V the product of the $2s$ copies of X at the places i or $n+i$ with $i \in \mathrm{orbit}_1(\sigma)$ and W the product of the remaining $2n - 2s$ copies of

X. By construction $\alpha = p_V^* \alpha'$ and $\beta = p_W^* \beta'$. Moreover β' is a zero-cycle on W. Using Lemmas 5.5.2 and 5.5.3 (applied to $p_{1,n+1} : V \to X \times X$ and the morphism $W \to \operatorname{Spec} k$) we can rewrite the summand corresponding to σ as

$$\operatorname{sgn}(\sigma) \deg(\beta')(p \circ f)^{s-1} \circ p = \operatorname{sgn}(\sigma) \deg(\beta') f^{s-1}$$

since $p \circ f = f \circ p = f$. If $s = n$, i.e., σ is an n-cycle, the corresponding summand is $(-1)^{n-1} f^{n-1}$. For $s < n$ we obtain monomials in f of lower degree. Hence there exists a polynomial $G(T) \in \mathbb{Q}[T]$ with leading term $(-1)^{n-1}(n-1)! \, T^{n-1}$ such that $G(f) = 0$. This proves part (a).

For part (b), note that if f is numerically equivalent to zero, the zero-cycle β' is numerically equivalent to zero for all σ of order less than n. Hence in this case only the leading term of the polynomial G survives, and we obtain $f^{n-1} = 0$.

Remarks. 1) In order to see what is going on in the proof of a), the reader is invited to write it out for $n = 3$.
2) With a bit more work, one can relate the polynomial $G(T)$ to the characteristic polynomial of f on $H^*(M)$; see [**Jann07**, Theorem 6.4.12].

5.6. Applications and Conjectures

Consider the functors

$$\begin{array}{ccccc}
\mathsf{Mot}_{\mathrm{rat}}(k) & \to & \mathsf{Mot}_{\mathrm{hom}}(k) & \to & \mathsf{Mot}_{\mathrm{num}}(k) \\
M = (X, p, m) & \mapsto & M_{\mathrm{hom}} = (X, p_{\mathrm{hom}}, m) & \mapsto & M_{\mathrm{num}} = (X, p_{\mathrm{num}}, m)
\end{array} \tag{34}$$

The forgetful functor $\mathsf{Mot}_{\mathrm{rat}}(k) \to \mathsf{Mot}_{\mathrm{hom}}(k)$ is *not* faithful, i.e.

$$\operatorname{Hom}_{\mathsf{Mot}_{\mathrm{rat}}(k)}(M, N) \to \operatorname{Hom}_{\mathsf{Mot}_{\mathrm{hom}}(k)}(M_{\mathrm{hom}}, N_{\mathrm{hom}})$$

is not injective in general, as is shown by the following example.

EXAMPLE 5.6.1. A cycle $Z \in Z^i(X)$ of codimension i defines a correspondence $Z \in \mathsf{Corr}^i(\mathrm{pt}, X)$ and hence a morphism $\mathbf{1} = (\mathrm{pt}, \mathrm{id}, 0) \to h_{\mathrm{rat}}(X)(i) = (X, \mathrm{id}, i)$ in $\mathsf{Mot}_{\mathrm{rat}}(k)$ and a morphism $f : \mathbf{L}^{\otimes i} \to \mathrm{ch}(X)$. If Z is homologically equivalent to zero but not rationally equivalent to zero, then $f \neq 0$ but $f_{\mathrm{hom}} = 0$.

DEFINITION 5.6.2. Let M be a Chow motive, and let M_{hom} be the image of M under homological equivalence. The Chow motive M is a *phantom motive* if $M \neq 0$, but $M_{\mathrm{hom}} = 0$ in $\mathsf{Mot}_{\mathrm{hom}}(k)$. (Equivalently, M is a phantom motive if $M \neq 0$ but $H(M) = 0$.)

Although the functor $\mathsf{Mot}_{\mathrm{rat}}(k) \to \mathsf{Mot}_{\mathrm{hom}}(k)$ is not faithful, one expects that phantom motives do not exist. This is the case if M is finite dimensional:

THEOREM 5.6.3. *Let $M = (X, p, m)$ be a finite dimensional Chow motive. Suppose that p is numerically trivial. Then $M = 0$, i.e. $p = 0$.*

Proof: We have a decomposition $M = M_+ \oplus M_-$ with projectors $\mathrm{id}_{M_\pm} = p_\pm$ and $p_\pm = p_\pm \circ p$. By assumption $\mathrm{id}_{M_\pm}$ is numerically trivial. Hence, by Theorem 5.5.1, they are nilpotent and hence trivial. $\qquad\square$

COROLLARY 5.6.4. *Suppose that $M = (X, p, m)$ is a finite dimensional Chow motive. Then M is not a phantom motive.*

Proof: If the projector p is homologically trivial, it is numerically trivial. $\qquad\square$

COROLLARY 5.6.5. *For a finite dimensional motive M, we have* $\dim M = \dim H(M)$.

Proof: It suffices to see this if M is evenly or oddly finite dimensional. For M evenly finite dimensional, the argument goes as follows. From Cor. 4.3.4 it follows that $\dim M \geq \dim H(M) = \dim H^+(M)$. Set $n = \dim M$. Then $\bigwedge^n M \neq 0$. Hence, by the previous corollary, $H(\bigwedge^n M) \neq 0$. But $H^+(\bigwedge^n M) = \bigwedge^n H^+(M)$ since M is evenly finite dimensional (again by Cor. 4.3.4) so $\dim H(M) \geq n = \dim M$ and hence we have equality. $\square$

Consider the ideal

$$J_M = \mathrm{Ker}\{\mathrm{End}_{\mathrm{Mot}_{\mathrm{rat}}(k)}(M) \to \mathrm{End}_{\mathrm{Mot}_{\mathrm{num}}(k)}(M_{\mathrm{num}})\}.$$

Finite-dimensionality implies that the ideal J_M is a nil-ideal, i.e., every $f \in J_M$ is nilpotent:

COROLLARY 5.6.6. *Let M be a finite dimensional motive. An endomorphism f of M which is numerically (or cohomologically) trivial is nilpotent. (Moreover, the order of nilpotency is uniformly bounded.)*

Proof: Decompose $M = M_+ \oplus M_-$; let $p_\pm$ the projectors to $M_\pm$ so that $\mathrm{id}_M = p_+ + p_-$ and we have a decomposition

$$
\begin{aligned}
f &= (p_+ + p_-) \circ f \circ (p_+ + p_-) \\
  &= f_+ + f_- + f_{\mathrm{mix}}, \quad f_+ = p_+ \circ f \circ p_+,\ f_- = p_- \circ f \circ p_-,\ f_{\mathrm{mix}} = p_+ \circ f \circ p_-,
\end{aligned}
$$

so that $f_\pm$ are endomorphisms of $M_\pm$ while f_{mix} is a morphism which does not preserve parity and hence, by Prop. 5.3.1 and Lemma 5.2.3, is smash nilpotent, say of order R (with R independent of f). So, by Prop. 5.2.5, if we expand out f^n, we get a sum of monomials in which f_{mix} does not appear more than R times. The powers of f_{mix} are separated by powers of f_+ of f_- and so a typical surviving monomial reads

$$f_\pm^{k_1} \circ f_{\mathrm{mix}}^{\ell_1} \circ f_\pm^{k_2} \circ f_{\mathrm{mix}}^{\ell_2} \circ \cdots \circ f_{\mathrm{mix}}^{\ell_r} \circ f_\pm^{k_{r+1}} \tag{35}$$

where

$$r \leq \ell_1 + \cdots \ell_r \leq R - 1.$$

On the other hand, by Theorem 5.5.1, the assumption that f and hence $f_\pm$ is numerically trivial implies that they are nilpotent, say $f_\pm^S = 0$ and this implies that if $n \geq SR$, at least one k_j must be $\geq S$ and so (35) vanishes in this case. Indeed, if all $k_j \leq S - 1$, we would have

$$n = \sum_{j=1}^{r+1} k_j + \sum_{j=1}^{r} \ell_j \leq (r+1)(S-1) + (R-1) \leq R(S-1) + R - 1 = RS - 1.$$

Summarizing, for $n \geq RS$ we have $f^n = 0$. $\square$

COROLLARY 5.6.7. *If M is a finite dimensional Chow motive, then the ideal J_M is nilpotent.*

Proof: By Corollary 5.6.6 J_M is a nil-ideal and the order of nilpotency is uniformly bounded. Hence, by a theorem of Nagata-Higman [**Andr-Ka-O'S**, 7.2.8] the ideal J_M is nilpotent. $\square$

CONJECTURE 5.6.8 (**Conjecture of Kimura-O'Sullivan**). Every Chow motive is finite dimensional.

This is related to two further conjectures:

CONJECTURE 5.6.9 (**Conjectures $N(M)$ and $N'(M)$**). The ideal J_M is nilpotent, respectively a nil-ideal.

In particular, taking $M = \mathsf{ch}(X)$ we have the **Conjecture $N(X)$**: the ideal

$$J(X) = \mathrm{Ker}\{\mathrm{Corr}^0_{\mathrm{rat}}(X, X) \to \mathrm{Corr}^0_{\mathrm{num}}(X, X)\}$$

is nilpotent.

Indeed, by Corollary 5.6.4 and Corollary 5.6.7 we have:

> Conjecture of Kimura-O'Sullivan $\Longrightarrow$ Conjecture $N(M)$
> and the absence of phantom motives.

Recall Voevodsky's conjecture 1.2.21, stating that any $f : M \to M$ which is numerically trivial is smash nilpotent and hence nilpotent (Theorem 5.2.4). Hence:

> Voevodsky's Conjecture $\Longrightarrow$ Conjecture $N'(M)$.

One can in fact show that Voevodsky's conjecture implies the Kimura-O'Sullivan Conjecture, which is a priori stronger than $N'(M)$. To explain this, we need a weak version of the standard conjecture $C(X)$ which states that the Künneth projectors are algebraic. Recall from § 4.5 that the *sign conjecture $S(X)$* states that the combined sum π_X^+ of all the even Künneth components and the combined sum π_X^- of all the odd Künneth components are algebraic.

The crucial results are as follows.

LEMMA 5.6.10 ([**Jann07**, Lemma 6.4.1]). *Let $M \in \mathsf{Mot}_{\mathrm{rat}}(k)$ and suppose that J_M is a nil-ideal, i.e. every $f \in J_M$ is nilpotent. Let M_{num} be the image of M in $\mathsf{Mot}_{\mathrm{num}}(k)$, as in (34). Then we have*

    (1) *If $M_{\mathrm{num}} = 0$, then $M = 0$;*

    (2) *Any idempotent in $\mathrm{End}(M_{\mathrm{num}})$ can be lifted to an idempotent in $\mathrm{End}(M)$, and any two such liftings are conjugate by a unit of $\mathrm{End}(M)$ lying above the identity of $\mathrm{End}(\bar{M})$.*

    (3) *If the image of $f \in \mathrm{End}(M)$ in $\mathrm{End}(M_{\mathrm{num}})$ is invertible, then so is f.*

Proof: For (1), note that If id_M maps to zero in $\mathrm{End}(M_{\mathrm{num}})$ it is nilpotent, hence zero.

Properties (2) and (3) hold for any surjection $A \twoheadrightarrow \bar{A} = A/I$ where A is a (not necessarily commutative) ring with unit, and I is a (two-sided) nil-ideal. We shall prove them in this general setting.

For (3), it suffices to assume that the element $a = 1 - j \in A$ maps to $1 \in \bar{A}$. But then j is nilpotent, say $j^{r+1} = 0$ and thus $1 + j + \cdots + j^r$ is an inverse for a. As for (2), if $\bar{e}$ is idempotent in $\bar{A}$ and a is any lift in A, then $(a - a^2)^N = 0$ for some $N > 0$, and it follows easily (i.e see the properties (1)–(4) in the proof of Theorem 5.4.4) that $\tilde{e} = (1 - (1 - a)^N)^N$ is an idempotent lifting $\bar{e}$. If e and e' are idempotents of A lying above $\bar{e}$, then $u = e'e + (1 - e')(1 - e)$ lies above $1 \in \bar{A}$. Thus u is invertible, and the equality $e'u = e'e = ue$ shows that $e' = ueu^{-1}$. $\qquad\square$

PROPOSITION 5.6.11 ([**Jann07**, Cor. 6.4.9]). *Let X be a smooth projective variety. Then the following statements are equivalent:*

    (1) *the Chow motive $\mathsf{ch}(X)$ is finite-dimensional.*

    (2) *$S(X)$ holds, and $N(X^n)$ holds for all $n \geq 1$.*

(3) $S(X)$ *holds, and* $N'(X^n)$ *holds for all* $n \geq 1$.

Proof: Write $M = \mathsf{ch}(X) = M_+ \oplus M_-$. Then $H(M_+)$ (resp. $H(M_-)$) is the even (resp. odd) degree part of the cohomology $H(M)$). Therefore, modulo homological equivalence, $p_\pm = \pi_\pm^X$. and so (1) implies $S(X)$. By Corollary 5.6.7 (1) implies that J_M is nilpotent, i.e. $N(X)$ holds. Since (1) also implies finite-dimensionality of $h(X^n) = h(X)^{\otimes n}$, for all $n \geq 1$, (1) implies (2).

(2) $\Rightarrow$ (3) is trivial.

(3) $\Rightarrow$ (1): If $S(X)$ holds, the $\pi_X^\pm$ are algebraic projectors and if $N'(X)$ holds these lift to orthogonal projectors $\tilde{\pi}_+$ and $\tilde{\pi}_-$ with $\tilde{\pi}_+ + \tilde{\pi}_- = \mathrm{id}$ (using Lemma 5.6.10, lift π_+ to a projector $\tilde{\pi}_+$ and let $\tilde{\pi}_- = \mathrm{id} - \tilde{\pi}_+$). Let $M_\pm = (X, \tilde{\pi}_\pm, 0)$. Then $M = M_+ \oplus M_-$, and for $b_\pm = \dim H(M_\pm)$ one has $\bigwedge^{b_++1} M_+ = 0 = \mathrm{Sym}^{b_-+1}(M_-)$ modulo homological equivalence, and hence also modulo numerical equivalence. By Lemma 5.6.10 (1) and $N'(X^n)$, for $n = b_+ + 1$ and $n = b_- + 1$, one concludes that this vanishing also holds modulo rational equivalence, i.e., we obtain (1). $\square$

Now we can show:

PROPOSITION 5.6.12. *Voevodsky's Conjecture implies the conjecture of Kimura-O'Sullivan.*

Proof: Recall that Voevodsky's conjecture implies the standard conjecture $D(X)$ (postulating $\sim_{\mathrm{hom}} = \sim_{\mathrm{num}}$) (see the lines after the statement of Voevodsky's Conjecture 1.2.21). By the discussion in § 3.1.4 we have $D(X) \implies A(X, L)$ and if this holds for all smooth projective X, it implies $B(X)$. But $B(X) \implies C(X)$ and so in particular $S(X)$ holds for all $X \in \mathsf{SmProj}(k)$. Moreover, Voevodsky's conjecture implies conjecture $N'(X)$ (see the previous page). Hence also $N'(X)$ holds for all $X \in \mathsf{SmProj}(k)$ and $\mathsf{ch}(X)$ is finite dimensional by Prop. 5.6.11. $\square$

We conclude this chapter by mentioning two further important consequences of finite dimensionality. The relevance of these results will be clarified in the next two chapters where also some auxiliary results will be explained.

Let us first recall (§ A-3):

$$\boxed{\text{Bloch Conjecture for surfaces: } H^2(S)_{\mathrm{trans}} = 0 \implies T(S) = 0}$$

Here S is a smooth projective surface defined over an algebraically closed field k. The assumption $H^2(S)_{\mathrm{trans}} = 0$ means that all of $H^2(S)$ comes from algebraic cycles (so, if $k = \mathbb{C}$ this is equivalent to $p_g = 0$). Recall (see formula (10)) also that

$$T(X) := \mathrm{Ker}\left[\mathsf{CH}_0^{\mathrm{hom}}(X) \xrightarrow{\mathrm{alb}_X} \mathrm{Alb}(X)\right].$$

PROPOSITION 5.6.13. *The Bloch Conjecture holds if* $\mathsf{ch}(S)$ *is finite dimensional.*

Proof: In § 6.3.2 we shall see that we can unconditionally construct a Chow motive $t^2(S) = (S, p, 0)$ which is a direct summand of $\mathsf{ch}(S)$ with the following two properties:

- its cohomology is $H^2(S)_{\mathrm{trans}}$;
- its Chow group is $T(S) \otimes \mathbb{Q}$.

Since $\mathsf{ch}(S)$ is assumed to be finite dimensional, by Cor. 5.4.6 any direct summand is. In particular $t^2(S)$ is finite dimensional. But by assumption its cohomology $H^2(S)_{\mathrm{trans}}$ is trivial from which it easily follows that p is homologically equivalent to zero and hence by Theorem 5.5.1 the projector p is nilpotent and therefore $p = 0$.

It follows that $t^2(S) = 0$ an hence its Chow group $T(S) \otimes \mathbb{Q}$ is trivial. It follows that $T(S)$ can only be torsion. However, Roitman's theorem [**Roi**] states that the albanese map is an isomorphism on the torsion points of $T(S)$ and so $T(S) = 0$.[1] $\square$

---

[1] See [**Kimu**, prop. 7.6 and Cor. 7.7] for another proof which does not use the existence of $t^2(S)$.

CHAPTER 6

Chow-Künneth Decomposition; the Picard and Albanese Motive

In this Chapter —as before — k is a field and $\mathsf{SmProj}(k)$ is the category of smooth projective varieties defined over k, $\mathsf{Mot}_{\mathrm{rat}}(k)$ the category of Chow-motives over k and for $X, Y \in \mathsf{SmProj}(k)$ the k-correspondences from X to Y with rational coefficients are denoted by $\mathsf{Corr}(X, Y)$.

We choose, once and for all, a Weil-cohomology theory $H(-)$, for instance étale cohomology $H_{\mathrm{\acute{e}t}}(X_{\bar{k}}, \mathbb{Q}_\ell)$ where $\ell \neq \mathrm{char}(k)$.[1]

6.1. The Künneth and the Chow-Künneth Decomposition

6.1.1. Künneth Conjecture.
Recall from § 3.1.1 that if $X = X_d$ and $\Delta(X) \subset X \times X$ the diagonal of X then

$$\gamma_{X \times X}(\Delta(X)) = \sum_{i=0}^{2d} \Delta_i^{\mathrm{topo}}(X) \in H^{2d}(X \times X) = \bigoplus_{i=0}^{2d} H^{2d-i}(X) \otimes H^i(X).$$

Recall the notation for the topological cycle class $\Delta_i^{\mathrm{topo}} \in H^{2d-i}(X) \otimes H^i(X)$, the i–th *Künneth component* of the diagonal. Using the isomorphism

$$H^{2d-i}(X) \otimes H^i(X) \cong H^i(X)^\vee \otimes H^i(X) \cong \mathrm{Hom}(H^i(X), H^i(X))$$

induced by Poincaré duality, we can view Δ_i^{topo} as a map $H^*(X) \to H^*(X)$ that acts as the identity on $H^i(X)$ and as zero on the other cohomology groups. Then (see § 3.1.1) the *Künneth conjecture* $C(X)$ is the assumption that there exist *algebraic* cycles $\Delta_i \in \mathsf{CH}^d(X \times X)$ such that $\gamma_{X \times X}(\Delta_i) = \Delta_i^{\mathrm{topo}}$ $(i = 0, \ldots, 2d)$.

6.1.2. Main Definitions.

DEFINITION 6.1.1. Let $X = X_d \in \mathsf{SmProj}(k)$. We say that X admits a *Chow-Künneth decomposition* (C-K decomposition for short) if there exist $p_i(X) \in \mathsf{CH}^d(X \times X)_{\mathbb{Q}} = \mathsf{Corr}^0(X, X)$ for $0 \leq i \leq 2d$ such that

(1) $\sum_{i=0}^{2d} p_i(X) = \Delta(X)$

(2) $p_i(X) \circ p_j(X) = \begin{cases} 0 & j \neq i \\ p_i(X) & j = i \end{cases}$

(3) $\gamma_{X \times X}(p_i(X)) = \Delta_i^{\mathrm{topo}}$.

Remarks. (1) So the $p_i(X)$ are *projectors, orthogonal* to each other, *lifting* the Künneth components and they are *idempotents* in $\mathsf{Corr}(X, X)$.

---

[1] We like to stress however that the constructions below are independent of the choice of the Weil cohomology theory (see Remark 6.2.2 3).

(2) If such projectors $p_i(X)$ exist, we put $\mathrm{ch}^i(X) = (X, p_i(X), 0)$. We then have a decomposition

$$\mathrm{ch}(X) = \bigoplus_{i=0}^{2d} \mathrm{ch}^i(X)$$

and call $\mathrm{ch}^i(X)$ the i–th *Chow-Künneth motive* of X.

(3) If X admits a k–rational point [2] $e \in X(k)$, take $p_0(X) = e \times X$ and $p_{2d}(X) = X \times e$. These are projectors, orthogonal to each other, lifting Δ_0^{topo} and $\Delta_{2d}^{\mathrm{topo}}$ respectively. They are the obvious candidates for the trivial part of the Chow-Künneth decomposition.

6.1.3. Chow-Künneth Conjecture. The following Conjecture is due to the first author [**Mur93**, Part I].

CONJECTURE 6.1.2 (Conjecture $CK(X)$). Every smooth projective variety admits a Chow-Künneth decomposition.

EXAMPLE 6.1.3. Let C be a smooth, projective curve that has a k–rational point $e \in C(k)$. Take $p_0(C) = e \times C$, $p_2(C) = C \times e$ and $p_1(C) = \Delta(C) - p_0(C) - p_2(C)$. These projectors give a Chow-Künneth decomposition

$$\mathrm{ch}(C) = \mathrm{ch}^0(C) \oplus \mathrm{ch}^1(C) \oplus \mathrm{ch}^2(C)$$

with $\mathrm{ch}^i(C) = (C, p_i(C), 0)$ $(i = 0, 1, 2.)$ See § 2.7.

Remarks.   (1) Clearly the Chow-Künneth conjecture $CK(X)$ implies the Künneth conjecture $C(X)$.

(2) We expect that the projectors can be chosen such that

$$p_{2d-i}(X) = {}^{\mathsf{T}}p_i(X).$$

If this is the case, we say that the C-K decomposition is *self-dual.*

The Kimura-O'Sullivan Conjecture 5.6.8 has an important consequence:

PROPOSITION 6.1.4. *If X has finite dimensional Chow motive, the Künneth conjecture $C(X)$ implies the Chow-Künneth conjecture $CK(X)$.*

Proof: By Prop. 5.6.11 if $\mathrm{ch}(X)$ is finite dimensional we know that conjecture $N(X)$ (5.6.9) is true. Now [**Jann94**, Lemma 5.4] states that $N(X)$ together with $C(X)$ implies $CK(X)$.

6.1.4. Further Remarks. The Künneth components $\Delta_i^{\mathrm{topo}}(X)$ are unique, but the projectors $p_i(X)$ are *not* unique as cycle classes! For instance, on a curve C the $p_i(C)$ depend on the choice of a k–rational point $e \in C(k)$. However we expect (conjecturally) that the Chow-Künneth motives $\mathrm{ch}^i(X)$ are unique up to a so called *natural isomorphism:*

DEFINITION 6.1.5. A morphism $f : M = (X, p, 0) \to M' = (X, p', 0)$ of motives is a *natural isomorphism* if it is an isomorphism and f is induced by the identity on the underlying variety X, i.e., if $f = p' \circ p$. In this case we say that M and M' are naturally isomorphic. Note that if $p \circ p' \circ p = p$ and $p' \circ p \circ p' = p'$, then $f = p' \circ p$ and $g = p \circ p'$ are natural isomorphisms which are moreover inverse to each other.

---

[2]If X does not admit a k–rational point we take a k-rational positive 0-cycle $\mathfrak{A}$, of degree n say, and we put $p_0(X) = \frac{1}{n}(\mathfrak{A} \times X)$ and $p_{2d}(X) = \frac{1}{n}(X \times \mathfrak{A})$.

For instance, if in $\mathsf{ch}^0(X) = (X, p_0 = e \times X, 0)$ we take another k–rational point e' and $p_0' = e' \times X$ then we have indeed (as one checks immediately) $p_0 \circ p_0' \circ p_0 = p_0$ and $p_0' \circ p_0 \circ p_0' = p_0'$, and the same result holds for $p_{2d} = X \times e$ and $p_{2d}' = X \times e'$. Moreover if X is a curve and we put $p_1 = \mathrm{id} - p_0 - p_2$, $p_1' = \mathrm{id} - p_0' - p_2'$ then we also have $p_1 \circ p_1' \circ p_1 = p_1$ and $p_1' \circ p_1 \circ p_1' = p_1'$.

We shall see that the conjectures stated in Chapter 7 imply that if M and M' in $\mathsf{Mot}_{\mathrm{rat}}(k)$ are as above and $p = p'$ modulo homological equivalence (i.e., with obvious notation $M_{\mathrm{hom}} = M'_{\mathrm{hom}}$) then M and M' are naturally isomorphic in $\mathsf{Mot}_{\mathrm{rat}}(k)$ via $f = p' \circ p : M \to M'$ and $g = p \circ p' : M' \to M$; see Corollary 7.5.8.

6.1.5. Examples Where $CK(X)$ is Known. [3]

(1) Curves (trivial; see 6.1.3)
(2) Surfaces ([**Mur90**], see § 6.3 below)
(3) Note that if X and Y both admit a C-K decomposition, then $CK(X \times Y)$ is true since one can take (see also § 7.4.1)

$$p_i(X \times Y) = \sum_{r+s=i} p_r(X) \times p_s(Y)$$

and we have

$$\mathsf{ch}(X \times Y) = \bigoplus_i (\bigoplus_{r+s=i} \mathsf{ch}^r(X) \otimes \mathsf{ch}^s(Y)).$$

Hence $CK(X)$ is true for products of curves and surfaces.

(4) Conjecture $CK(X)$ holds for abelian varieties. This result is already implicit (but not formulated in the context of C-K decompositions) in a paper by Shermenev [**Serm**]. For more modern proofs (for an abelian scheme A over a base scheme S) see [**Den-Mu, Kü93**]. These modern proofs are based on the theory of the so-called *Fourier transform* on abelian varieties developed by Mukai and Beauville [**Muk, Beau83**].
(5) Uniruled threefolds [**Ang-MüS98**]. Also for threefolds satisfying certain conditions on the transcendental part of $H^2(X)$ ([**Ang-MüS00**], and recently [**MüS-Sa**]).
(6) Elliptic modular varieties (certain Kuga-Satake varieties): [**Gor-Ha-Mu**].

For the following two examples, the projectors $p_i(X)$ can be constructed because there is transcendental cohomology only in the middle dimension (see Appendix C).

(8) Complete intersections in projective space.
(9) Calabi-Yau threefolds.

6.1.6. Open Question.

Conjecture $C(X)$ is known for varieties defined over a finite field $\mathbb{F}_q$ ([**Katz-Me**], the result is obtained using Deligne's proof of the Riemann-Weil conjecture). However, to our knowledge $CK(X)$ is *not* known for X defined over a finite field. In fact the $\Delta_i(X)$ are given by polynomials in the powers of the Frobenius with rational coefficients. These polynomials give the Künneth components for homological equivalence (loc. cit., see also [**Lev08**, p. 210]). Are the $p_i(X)$ given by the same polynomials?

---

[3]This is not an exhaustive list of all known examples.

6.2. Picard and Albanese Motives

Let $X = X_d \in \mathsf{SmProj}(k)$, and let $e \in X(k)$ be a k–rational point.[4] We have the two projectors $p_0(X) = e \times X$ and $p_{2d}(X) = X \times e$ (see Chapter 2 § 2.3(iii)), orthogonal to each other. It is possible to construct two more projectors $p_1(X)$ and $p_{2d-1}(X)$ which lift the corresponding Künneth components.

THEOREM 6.2.1 ([**Mur90**]). *Let $X = X_d \in \mathsf{SmProj}(k)$ (any $d > 0$). There exist $p_1(X)$ and $p_{2d-1}(X)$ in $\mathsf{CH}^d(X \times X)_{\mathbb{Q}}$ with the following properties.*

(i) *$p_1(X)$ and $p_{2d-1}(X)$ are projectors, orthogonal to each other and orthogonal to $p_0(X)$ and $p_{2d}(X)$. The projectors $p_1(X)$ and $p_{2d-1}(X)$ lift the Künneth components Δ_1^{topo} and $\Delta_{2d-1}^{\mathrm{topo}}$ and satisfy the relation ${}^{\mathsf{T}}p_1(X) = p_{2d-1}(X)$.*

(ii) *Let $\mathsf{ch}^1(X) = (X, p_1(X), 0) \in \mathsf{Mot}_{\mathrm{rat}}(k)$. Then*

$$
H^i(\mathsf{ch}^1(X)) = \begin{cases} 0 & i \neq 1 \\ H^1(X) & i = 1 \end{cases}
$$

$$
\mathsf{CH}^i(\mathsf{ch}^1(X)) = \begin{cases} 0 & i \neq 1 \\ (\mathrm{Pic}_X^0)_{\mathrm{red}}(k)_{\mathbb{Q}} & i = 1 \end{cases}
$$

where Pic_X is the Picard scheme of X, Pic_X^0 the connected component of the identity and $(\mathrm{Pic}_X^0)_{\mathrm{red}}$ the underlying reduced scheme, i.e., the classical Picard variety P_X of X. (So we take the group of k–rational points on the Picard variety and tensor this group with $\mathbb{Q}$.)

More precisely, $p_1(X)$ acts as the identity on $H^1(X)$ and on the Picard variety $P_X = (\mathrm{Pic}_X^0)_{\mathrm{red}}$. Hence $p_1(X)$ also acts as the identity on $\mathsf{CH}^1_{\mathrm{alg}}(X)_{\mathbb{Q}}$.

(iii) *Let $\mathsf{ch}^{2d-1}(X) = (X, p_{2d-1}(X), 0)$. Then*

$$
H^i(\mathsf{ch}^{2d-1}(X)) = \begin{cases} 0 & i \neq 2d - 1 \\ H^{2d-1}(X) & i = 2d - 1 \end{cases}
$$

$$
\mathsf{CH}^i(\mathsf{ch}^{2d-1}(X)) = \begin{cases} 0 & i \neq d \\ \mathrm{Alb}_X(k)_{\mathbb{Q}} & i = d \end{cases}
$$

where Alb_X is the Albanese variety A_X of X. More precisely, $p_{2d-1}(X)$ acts as the identity on $H^{2d-1}(X)$ and on the Albanese variety Alb_X.

Remarks 6.2.2. (1) The above statements about the actions of $p_1(X)$ and $p_{2d-1}(X)$ remain true after an arbitrary base extension K of k. This follows from the proof of the theorem (see below) and it depends on the fact that the Picard and Albanese variety are already defined over the same field of definition as the variety X, i.e., over k.

(2) Because of the properties (ii) and (iii) we call $\mathsf{ch}^1(X)$ the *Picard motive* of X and $\mathsf{ch}^{2d-1}(X)$ the *Albanese motive*.

(3) A priori (ii) and (iii) seem, as far as the cohomology is concerned, to depend on the choice of the Weil cohomology. However in fact we need only the independence of $H^1_{\text{ét}}(X_{\bar{k}}, \mathbb{Q}_\ell)$, $\ell \neq p$), which gives by duality also the independence of $H^{2d-1}_{\text{ét}}(X_{\bar{k}}, \mathbb{Q}_\ell)$. The required independence is a consequence of the fact [**Mil80**, p. Prop. 4.11] that $H^1_{\text{ét}}(X_{\bar{k}}, \mathbb{Z}/\ell^\nu) \simeq$

---

[4]See footnote 2.

$P_X(X)[\ell^\nu]$, the group of torsion points of order ℓ^ν inside the reduced Picard scheme.

Proof of Theorem 6.2.1. (see also [**Mur90**] and [**Scholl**, p. 176])

First we give the **idea of the proof.**Let L_X be the Lefschetz operator of X and Λ_X the lambda operator (see § 3.1.2). The standard conjecture $B(X)$ predicts that Λ_X is algebraic. This is not yet known, but the iterated operator

$$\Lambda_X^{d-1} = \Lambda_X \circ \Lambda_X \circ \ldots \circ \Lambda_X$$

is algebraic and is a *divisor class* [**Klei68**] The proof of the existence of $p_1(X)$ and $p_{2d-1}(X)$ is an *algebraic refinement* of this result that uses the theory of the Picard and Albanese variety as recalled in Appendix A-2. This may not seem apparent at first sight; see however Remarks 6.2.5 (1), in which we explain that Λ_X^{d-1} is identified with the correspondence induced by the isogeny β constructed below. This isogeny in turn is used to construct the two projectors p_1 and p_{2d-1}.

We now proceed to the actual **proof.**

(a) Construction of $p_1(X)$ and $p_{2d-1}(X)$. We have fixed a point $e_X \in X_d \subset \mathbb{P}^N$. Consider the family $H(\underline{t})$ ($\underline{t} = (t_1, \ldots, t_r) \in T$, a linear space) of hyperplanes through e_X and the family of curves (linear curve sections)

$$C(\underline{t}) = X \cap H(t_1) \cap \ldots \cap H(t_{d-1}) \overset{i(\underline{t})}{\hookrightarrow} X.$$

Take a fixed *smooth* curve

$$C = C(\underline{t_0}) \overset{i}{\hookrightarrow} X, \quad \underline{t_0} \in T(k).$$

Such a curve C is a *typical curve* (in Weil's terminology *courbe typique* [**Weil54**, p. 118](=Collected Papers, Vol 2, p. 150), i.e. C is a linear section of X and all points of C are smooth, both on X as well as on C. From such a curve we get morphisms of abelian varieties

$$\mathrm{pic}(i) : P_X \to \mathrm{J}(C), \quad \mathrm{alb}(i) : \mathrm{Alb}_C \to \mathrm{Alb}_X .$$

In the sequel we shall write

$$i^* = \mathrm{pic}(i), \quad \text{and } i_* = \mathrm{alb}(i)$$

by abuse of notation. Using the identification $P_C = \mathrm{Alb}_C$ we obtain the morphism

$$\alpha = i_* \circ i^* : P_X \to \mathrm{Alb}_X .$$

LEMMA 6.2.3. *The morphism* $\alpha : P_X \to \mathrm{Alb}_X$ *satisfies the following properties.*

(1) α *is an isogeny of abelian varieties;*
(2) α *does not depend on the choice of the point $\underline{t_0}$ used to construct the curve* $C = C(\underline{t_0})$;
(3) $\widehat{\alpha} = \alpha$.

Proof: We start with part (1). Since $(\mathrm{Pic}_X^0)_{\mathrm{red}}$ and Alb_X have the same dimension, it suffices to show that $\mathrm{Ker}(\alpha)$ is finite. By [**Weil54**, Thm. 7, Cor. 1, p. 127](=Collected Papers, vol. 2, p. 159), since C is a typical curve, the morphism $i^* : P_X \to \mathrm{J}(C) = P_C$ has finite kernel. Let P_X' be the image of i^* in $\mathrm{J}(C)$ and consider the canonical factorization

$$P_X \overset{i_1}{\longrightarrow} P_X' \overset{j}{\longrightarrow} \mathrm{J}(C).$$

Since i_1 is an isogeny, to see that $\alpha = i_* \circ i^*$, where $i_* = \mathrm{alb}(i)$, has finite kernel it suffices to show that $i_* \circ j$ has finite kernel. Now

$$i_* : \mathrm{J}(C) = \widehat{\mathrm{J}(C)} \to \mathrm{Alb}_X = A_X$$

is the dual of i^* and hence $i_* = \widehat{i^*} = \widehat{i_1} \circ \widehat{j}$ with $\widehat{i_1}$ again an isogeny. So, finally, it suffices to show that

$$\widehat{j} \circ j : P'_X \to \mathrm{J}(C) \simeq \widehat{\mathrm{J}(C)} \to \widehat{P'_X}$$

has finite kernel.

Now recall that the identification of the jacobian with its dual is obtained via the theta-divisor Θ:

$$\varphi_\theta : \mathrm{J}(C) \xrightarrow{\;\simeq\;} \widehat{\mathrm{J}(C)}$$
$$x \quad \mapsto \quad [\Theta_x - \Theta],$$

where Θ_x is the translate of the the theta-divisor by x and $[\Theta_x - \Theta]$ is the linear equivalence class of the divisor $\Theta_x - \Theta$. Now, quite generally, if $\lambda : A \to B$ is a homomorphism of abelian varieties, then $\hat{\lambda} : \hat{B} \to \hat{A}$ is given by $\lambda^*(D)$ for a divisor class D on B algebraically equivalent to zero [**La**, second half of p.124]; so in our case $\hat{j}(D) = j^*(D) = D \cdot P'_X$. Therefore, $\hat{j} \circ j$ is given as follows: starting with $y \in P'_X$ and setting $x = j(y)$, one has

$$(\hat{j} \circ j)(y) = (\Theta_x - \Theta) \cdot P'_X = (E_y - E) = \varphi_E(y),$$

where $E = \Theta \cdot P'_X$. Since Θ is ample on $\mathrm{J}(C)$, also E is ample on P'_X. It now follows that $\mathrm{Ker}(\hat{j} \circ j) = \mathrm{Ker}\,\varphi_E$ is finite upon applying the following general fact about abelian varieties [**La**, p. 85] or [**Mum74**, App I, p.57–58]: if A is an abelian variety and E is a divisor class on A the homomorphism φ_E has finite kernel if and only if E is ample.

Next we turn to the proof of part (2) of the Lemma. The map α does not depend on the choice of $\underline{t}_0$ in the construction of C since by a result of Chow (see [**La**, Thm. 5, p. 26]) there is no continuous family $\alpha(\underline{t}) : (\mathrm{Pic}^0_X)_{\mathrm{red}} \to \mathrm{Alb}_X$ of homomorphisms, and so $\alpha(\underline{t})$ is defined over k itself and hence $\alpha(\underline{t}) = \alpha(\underline{t}_0)$.[5]

Part (3) of the Lemma follows formally:

$$\widehat{\alpha} = \widehat{i_* \circ i^*} = \widehat{i^*} \circ \widehat{i_*} = i_* \circ i^* = \alpha. \qquad \square$$

Remarks.   (1) An alternative proof of part (1) of the Lemma is obtained via ℓ–adic cohomology $\ell \neq \mathrm{char}(k)$ and hard Lefschetz as follows. Hard Lefschetz states that we have an isomorphism

$$L_X^{d-1} : H^1_{\text{ét}}(X_{\overline{k}}, \mathbb{Q}_\ell) \xrightarrow{\;\sim\;} H^{2d-1}_{\text{ét}}(X_{\overline{k}}, \mathbb{Q}_\ell).$$

Now recall that for the ℓ^n–torsion points of the Picard variety we have ([**Mil80**, Cor. 4.18, p.131])

$$P_X[\ell^n] \xrightarrow{\;\simeq\;} H^1_{\text{ét}}(X_{\overline{k}}, \mathbb{Z}/\ell^n\mathbb{Z}). \tag{36}$$

Passing to the limit of this directed system, we get

$$T_\ell(P_X) \xrightarrow{\;\sim\;} H^1_{\text{ét}}(X_{\overline{k}}, \mathbb{Z}_\ell)$$

---

[5] In order to check that Chow's theorem applies we need to observe that $k(\underline{t})$ is primary extension of k since $k(\underline{t}) \cap \overline{k} = k$.

where $T_\ell(P_X) = \varprojlim_n P_X[\ell^n]$ is the ℓ–*adic Tate group* of P_X (and recall the definition of $H^1_{\text{ét}}(X_{\bar k}, \mathbb{Z}_\ell)$ from Chapter 1, Ex. 1.2.14 (2)). Similarly we have

$$T_\ell(A_X) \xrightarrow{\sim} H^{2d-1}_{\text{ét}}(X_{\bar k}, \mathbb{Z}_\ell).$$

Therefore hard Lefschetz translates into an isomorphism

$$L^{d-1}_X : T_\ell(P_X) \otimes_{\mathbb{Z}_\ell} \mathbb{Q}_\ell \xrightarrow{\sim} T_\ell(A_X) \otimes_{\mathbb{Z}_\ell} \mathbb{Q}_\ell$$

and since L^{d-1}_X corresponds to our homomorphism $\alpha : P_X \to A_X$ it follows that $(\operatorname{Ker}\alpha)^0 = 0$, where the superscript 0 denotes the connected component of the identity. Hence α has finite kernel and is an isogeny.
  (2) By Poincaré reducibility [**La**, Thm. 6, p. 28], [**Mum74**, Thm. 1, p. 160] we have $J(C) \simeq P'_X + (\operatorname{Ker}\hat{\jmath})^0$.
  (3) By Chow's theory of the Picard variety [**La**, Ch. 7, Thm. 12, p. 223] we have for a general linear curve section $i(\underline{t}) : C(\underline{t}) \hookrightarrow X$ that $(\operatorname{Pic}^0_X)_{\text{red}}$ is the $k(\underline{t})/k$-trace of the family $J(C(\underline{t}))$ and then

$$J(C(\underline{t})) \simeq i(\underline{t})^*(\operatorname{Pic}^0_X)_{\text{red}} + \operatorname{Ker} i(\underline{t})^0_*.$$

Corollary 6.2.4.   (1) *There exists an isogeny* $\beta : \operatorname{Alb}_X \to P_X$ *such that* $\alpha \circ \beta = m.\operatorname{id}_{\operatorname{Alb}_X}$ *for some integer* m;
  (2) $\beta = \hat{\beta}$;
  (3) $\beta \circ \alpha = m.\operatorname{id}_{P_X}$.

Proof: Immediate from the theory of abelian varieties [**La**, p. 29].   □

Remarks 6.2.5.   (1) The isogeny β induces the iterate Λ^{d-1}_X of the lambda operator.
  (2) As we have seen in Appendix A-2 (b), β determines a divisor class $D(\beta) := \Phi(\beta) \in \mathsf{CH}^1(X \times X) = \mathsf{Corr}^{1-d}(X, X)$ normalized by $D(\beta)(e) = 0$ and $^\mathsf{T}D(\beta)(e) = 0$. The relation $\beta = \hat{\beta}$ implies that $D(\beta) = {}^\mathsf{T}D(\beta)$.

Now define

$$p_1(X) := \frac{1}{m} D(\beta) \circ \Gamma_i \circ {}^\mathsf{T}\Gamma_i.$$

By abuse of notation we shall also write

$$p_1(X) = \frac{1}{m} D(\beta) \circ i_* \circ i^*.$$

To be precise $i^* = {}^\mathsf{T}\Gamma_i$ is a degree 0 correspondence from X to C, $i_* = \Gamma_i$ a correspondence from C to X of degree $d-1$ so that $p_1(X)$ is a self-correspondence of X of degree 0 as it should be. Introduce also the degree 0 correspondence $D(\beta)_C$ of C to X figuring as the first half of the decomposition of $p_1(X)$:

$$p_1(X) = \underbrace{\frac{1}{m} D(\beta) \circ i_*}_{D(\beta)_C} \circ i^*. \tag{37}$$

Informally, $p_1(X)$ is the restriction of $\frac{1}{m} D(\beta)$ to $C \times X$, viewed as a cycle on $X \times X$. Also note that $D(\beta)_C \in \mathsf{CH}^1(C \times X)_{\mathbb{Q}}$.

Lemma 6.2.6.   *The correspondence* $p_1(X)$ *is a projector on* X.

Proof: It suffices to show that

$$\frac{1}{m}D(\beta)\circ i_*\circ i^*\circ D(\beta)\circ i_* = D(\beta)\circ i_*.$$

This is a relation between *divisor classes* on $C \times X$. This translates, via the isomorphism Φ^{-1} from Appendix A, section A-2 (b), in a relation between homomorphisms of abelian varieties from $J(C)$ to P_X. More precisely, the divisor class $D(\beta)\circ i_*$ on the right hand side corresponds to $\beta\circ\mathrm{alb}(i) : J(C) \to P_X$. On the left hand side, consider first the divisor class $i^*\circ D(\beta)\circ i_* \in \mathsf{CH}^1(C\times C)$. By A-2 (c) this corresponds to the homomorphism

$$\mathrm{pic}(i)\circ\beta\circ\mathrm{alb}(i) : J(C) \to A_X \to P_X \to J(C).$$

Combining this with the divisor class $\frac{1}{m}D(\beta)\circ i_*$ and using the fact mentioned at the end of A-2 (c) we obtain on the left hand side the homomorphism

$$\frac{1}{m}\beta\circ\mathrm{alb}(i)\circ\mathrm{pic}(i)\circ\beta\circ\mathrm{alb}(i) : J(C) \to P_X.$$

Therefore in order to prove the relation of divisor classes stated above, we have to prove the following relation of homomorphisms from $J(C)$ to P_X:

$$\frac{1}{m}\beta\circ\mathrm{alb}(i)\circ\mathrm{pic}(i)\circ\beta\circ\mathrm{alb}(i) = \beta\circ\mathrm{alb}(i),$$

but since $\mathrm{alb}(i)\circ\mathrm{pic}(i)\circ\beta = \alpha\circ\beta = m.\,\mathrm{id}_{\mathrm{Alb}_X}$ this is true. $\square$

Note for later:

$$p_1(X) \text{ can be supported by a cycle on } C \times X. \tag{38}$$

Finally, put $p_{2d-1}(X) = {}^{\mathsf{T}}p_1(X)$. Clearly this is again a projector on X. This finishes part (i) of the theorem, except for the assertions on orthogonality (which will be proved later) and the lifting of the Künneth components, see below.

(b) Properties of $\mathrm{ch}^1(X) = (X, p_1(X), 0)$. We now turn to the proof of part (ii) of the theorem. First we look at the action of $p_1(X)$ on the Chow groups. This action factors in the following way:

$$\mathsf{CH}^p(X) \xrightarrow{i^*} \mathsf{CH}^p(C) \xrightarrow{i_*} \mathsf{CH}^{p+d-1}(X) \xrightarrow{D(\beta)_*} \mathsf{CH}^p(X).$$

For $Z \in \mathsf{CH}^p(X)$ with $p > 1$ we immediately get $p_{1,*}(Z) = 0$ since $i^*Z = 0$. For $p = 0$ we must look at $p_{1,*}[X] = n[X]$, $n \in \mathbb{Z}$. Composing with the homomorphism $e^* : \mathsf{CH}^0(X) \to \mathsf{CH}^0(X)$ (evaluation at e) and using Remark 6.2.5 (2), we obtain $n = 0$ since ${}^{\mathsf{T}}D(\beta)_*(e) = 0$.

Finally we turn to the case $p = 1$. For every $D \in \mathsf{CH}^1(X)$ we get $p_1(X)_*(D) = D(\beta)_C(C \cdot D)$, and this divisor is algebraically equivalent to zero because of the normalization $D(\beta)(e) = 0$ (recall that by construction $e \in X$ lies on the curve C!). Hence $p_1(X)_*(D) \in \mathsf{CH}^1_{\mathrm{alg}}(X)$. Now an element $D \in \mathsf{CH}^1_{\mathrm{alg}}(X)$ corresponds to a point $a \in P_X = \mathrm{Pic}^0_X(k)$, and $p_1(X)$ maps D to $\frac{1}{m}D(\beta)\circ i_*\circ i^*(D)$ which corresponds to the point $\frac{1}{m}\beta\circ\alpha(a) = a$ on the Picard variety P_X since $\beta\circ\alpha = m.\,\mathrm{id}_{P_X}$, hence $p_1(X)$ acts as the identity on P_X and on $\mathsf{CH}^1_{\mathrm{alg}}(X)$.

Remark 6.2.7. We are working with Chow groups with $\mathbb{Q}$–coefficients, but we could have done the construction of $p_1(X)$ and $p_{2d-1}(X)$ using $\mathsf{CH}^*(X \times X)[\frac{1}{m}]$, where m is the integer appearing via the isogenies α and β in the identity $\alpha\circ\beta = m.\,\mathrm{id}_{\mathrm{Alb}_X}$. Note that the integer m is *independent of the choice of a Weil cohomology theory* and also the action of $p_1(X)$ on P_X (resp. the action of $p_{2d-1}(X)$ on Alb_X)

is independent of the choice of the cohomology theory. Hence $p_1(X)$ and $p_{2d-1}(X)$ act on $\mathsf{CH}^*(X \times X)[\frac{1}{m}]$ and $H^*_{\text{ét}}(X_{\bar{k}}, \mathbb{Z}/\ell^\nu\mathbb{Z})$ provided that we choose a prime $\ell \neq p$ such that $(\ell, m) = 1$.

Next we turn to the cohomology. For the action on $H^j(X)$ with $j > 2$ we get zero, since i^* acts as zero. For $j = 2$, let $\sigma \in H^2(X)$. Then we have $p_1(X)_*(\sigma) = D(\beta)_C(i^*(\sigma))$ and $i^*(\sigma) \in H^2(C)$. Now either $i^*(\sigma) = 0$ or $i^*(\sigma) = \gamma_C(D)$ with $D \in \mathsf{CH}^1(C)_\mathbb{Q}$. However in the latter case we have already seen above that $D(\beta)_C(D) \in \mathsf{CH}^1_{\text{alg}}(X)_\mathbb{Q}$ since $D(\beta)_C(e) = 0$, hence its cohomology class is zero.

(c) Properties of $\mathsf{ch}^{2d-1}(X) = (X, p_{2d-1}(X), 0)$. Since $p_{2d-1}(X) = {}^\mathsf{T}p_1(X)$, the assertions in part (iii) of the theorem follow from those of (i) as far as cohomology is concerned.

So consider the action of $p_{2d-1}(X)$ on the Chow groups. Since by (38) the cycle $p_1(X)$ can be supported on $C \times X$, where C is the linear curve section, it follows that $p_{2d-1}(X)$ can be supported on $X \times C$ and hence $(p_{2d-1}(X))_*(Z)$ is supported on C for every $Z \in \mathsf{CH}^j(X)_\mathbb{Q}$. Consequently, $p_{2d-1}(X)$ acts as zero on $\mathsf{CH}^j(X)_\mathbb{Q}$ if $j \notin \{d-1, d\}$. For $Z \in \mathsf{CH}^{d-1}(X)_\mathbb{Q}$ we get $(p_{2d-1}(X))_*(Z) = n(Z) \cdot [C]$, so we must show $n(Z) = 0$. This is again done by evaluation at e, using the normalization ${}^\mathsf{T}D(\beta)(e) = 0$.

Finally let $Z \in \mathsf{CH}^d(X)_\mathbb{Q}$ and put $Z' := (p_{2d-1}(X))_*Z$. Since $p_{2d-1}(X)$ is orthogonal to $p_{2d}(X)$ it follows that

$$(p_{2d}(X))_*(Z') = (p_{2d}(X)_* \circ p_{2d-1}(X))_*(Z) = 0$$

on the one hand, but equals $(p_{2d}(X))_*(Z') = \deg(Z') \cdot Z'$ on the other hand so that $\deg Z' = 0$. Therefore it suffices to consider zero cycles of degree zero.

Since $p_1(X)$ operates as the identity on $(\text{Pic}^0_X)_{\text{red}}$, also $p_{2d-1}(X)$ operates as the identity on Alb_X. Hence $\text{Ker}(p_{2d-1}(X))|_{\mathsf{CH}^d(X)_\mathbb{Q}}$ is contained in the *Albanese kernel*

$$T(X)_\mathbb{Q} = \text{Ker}\{\mathsf{CH}^d_{\text{hom}}(X)_\mathbb{Q} \to \text{Alb}_X\}.$$

To show that $T(X)_\mathbb{Q} = \text{Ker}(p_{2d-1}(X))$ note that by Remark 6.2.5 (2)

$$p_{2d-1}(X) = {}^\mathsf{T}p_1(X) = \frac{1}{m}i_* i^* {}^\mathsf{T}D(\beta) = \frac{1}{m}i_* i^* D(\beta) = i_* \circ {}^\mathsf{T}D(\beta)_C.$$

By using the commutative diagram (where for simplicity of notation we omit to write the subscripts which indicate tensoring with $\mathbb{Q}$)

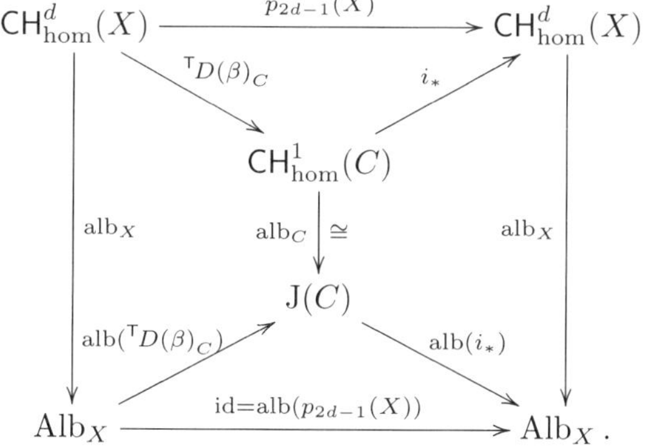

it follows that $\text{Ker}(p_{2d-1}(X)) = \text{Ker}(\text{alb}_X)$.

Remark 6.2.8. Note that the essential part in the above proof is our knowledge of the situation on curves; the proof boils down to the isomorphism $\mathrm{alb}_C : \mathrm{CH}^1_{\mathrm{hom}}(C) \xrightarrow{\sim} J(C)$.

(d) Orthogonality. This is the only part of Theorem 6.2.1 that remains to be proved. We leave it to the reader to show that $p_1(X)$ and $p_{2d-1}(X)$ are orthogonal to both p_0 and p_{2d} (see for instance [**Scholl**, p. 176-177]).

LEMMA 6.2.9. *If $d \geq 2$ then $p_{2d-1}(X) \circ p_1(X) = 0$.*

Proof: We have

$$p_{2d-1}(X) \circ p_1(X) = \frac{1}{m^2} i_* \circ i^* \circ {}^\mathsf{T}D(\beta) \circ D(\beta) \circ i_* \circ i^*.$$

If $d > 2$ then ${}^\mathsf{T}D(\beta) \circ D(\beta) \in \mathsf{Corr}^{2-2d}(X, X) = \mathsf{CH}^{2-d}(X \times X)$ vanishes. If $d = 2$ then ${}^\mathsf{T}D(\beta) \circ D(\beta) = n[X \times X]$ for some integer n. By evaluation at e we obtain $n = 0$, since $D(\beta)(e) = 0$. $\square$

LEMMA 6.2.10. *If $d > 2$ then $p_1(X) \circ p_{2d-1}(X) = 0$.*

Proof: We have

$$p_1(X) \circ p_{2d-1}(X) = \frac{1}{m^2} D(\beta) \circ i_* \circ i^* \circ i_* \circ i^* \circ {}^\mathsf{T}D(\beta).$$

Note that

$$i_* \circ i^* = p_{13,*}(({}^\mathsf{T}\Gamma_i \times X) \cdot (X \times \Gamma_i)) = \Delta(C) \subset X \times X.$$

Hence

$$i_* \circ i^* \circ i_* \circ i^* = p_{13,*}\{(\Delta(C) \times X) \cdot (X \times \Delta(C))\}$$

vanishes if $d > 2$ for dimension reasons (in fact the intersection product $(\Delta(C) \times X) \cdot (X \times \Delta(C))$ is zero). $\square$

END OF THE PROOF: *the case $d = 2$ (Surfaces)*
    In the case of a surface $X = S$ we have $p_3(S) \circ p_1(S) = 0$ but we can *not* conclude that $p_1(S) \circ p_3(S) = 0$. Indeed, let us calculate $p_1(S) \circ p_3(S)$ using

$$i_* \circ i^* \circ i_* \circ i^* = \Delta(A),$$

where $A = C \cdot C$ is a zero–cycle on S. If we write $A = \sum_i a_i$, we find that

$$p_1(S) \circ p_3(S) = \frac{1}{m^2} \sum_i {}^\mathsf{T}D(\beta)(a_i) \times D(\beta)(a_i)$$

is a sum of terms, each of which is a product of two divisors (and each of these divisors is algebraically equivalent to zero on S).
    So, we need to correct the $p_1(S)$ and $p_3(S)$ for $d = 2$. Let us denote (for $d = 2$ only!) the previously constructed projectors by $p_1(S)^{\mathrm{old}}$ and $p_3(S)^{\mathrm{old}}$, and define[6]

$$\left.\begin{array}{rcl} p_1(S)^{\mathrm{new}} & = & p_1(S)^{\mathrm{old}} - \frac{1}{2}p_1(S)^{\mathrm{old}} \circ p_3(S)^{\mathrm{old}} \\ p_3(S)^{\mathrm{new}} & = & p_3(S)^{\mathrm{old}} - \frac{1}{2}p_1(S)^{\mathrm{old}} \circ p_3(S)^{\mathrm{old}} \end{array}\right\} . \tag{39}$$

Then these new correspondences satisfy all the requirements of the theorem (the orthogonality is checked formally using the relation $p_3(S)^{\mathrm{old}} \circ p_1(S)^{\mathrm{old}} = 0$.)

---

[6]Following a suggestion of Scholl, we take the above correction of both $p_1(S)$ and $p_3(S)$ with a factor $\frac{1}{2}$ in order to keep the relation $p_3(S) = {}^\mathsf{T}p_1(S)$.

Remark 6.2.11. The motives $(S, p_1^{\mathrm{old}}(S), 0)$ and $\mathsf{ch}^1(S) = (S, p_1(S)^{\mathrm{new}}, 0)$ are naturally isomorphic (see Definition 6.1.5), as the morphisms

$$f = p_1(S)^{\mathrm{new}} \circ p_1(S)^{\mathrm{old}} : (S, p_1^{\mathrm{old}}(S), 0) \to \mathsf{ch}^1(S)$$

and

$$g = p_1(S)^{\mathrm{old}} \circ p_1(S)^{\mathrm{new}} : \mathsf{ch}^1(S) \to (S, p_1^{\mathrm{old}}(S), 0)$$

are inverse to each other. (One checks this by an easy computation using the identity $p_3(S)^{\mathrm{old}} \circ p_1(S)^{\mathrm{old}} = 0$.) Taking the transpose, we get the same result for $\mathsf{ch}^3(S)$ and $(S, p_3(S)^{\mathrm{old}}, 0)$.

6.2.1. On the Uniqueness of the Picard and Albanese Motives. The construction of $\mathsf{ch}^1(X)$ and $\mathsf{ch}^{2d-1}(X)$ depends on choices, like the choice of the polarization, the choice of the linear curve section $i : C \hookrightarrow X$ and in fact upon the entire method of construction. The projectors p_i themselves are certainly not unique, as is already clear in the case of curves where the p_i depend on the choice of a point $e \in X$. However the first author *conjectures* that the $\mathsf{ch}^i(X)$ are *unique up to natural isomorphism* in $\mathsf{Mot}_{\mathrm{rat}}(k)$ (recall here Definition 6.1.5). This is true in the case of curves (see subsection 6.1.4), and in general it would follow from the conjectures stated in the next Chapter 7; see Cor. 7.5.8.

In fact we have *unconditionally* that the motives $\mathsf{ch}^1(X)$ and $\mathsf{ch}^{2d-1}(X)$ are unique up to isomorphism due to the following theorem.

THEOREM 6.2.12. *The Picard motive* $\mathsf{ch}^1(X) = (X, p_1(X), 0)$ *from Theorem 6.2.1 is isomorphic to a direct summand of the motive* $\mathsf{ch}^1(C) = (C, p_1(C), 0)$, *where C is the linear section of X used in the construction of $p_1(X)$. It has finite dimension in the sense of Kimura–O'Sullivan. Moreover in* $\mathsf{Mot}_{\mathrm{rat}}(k)$ *the motive* $\mathsf{ch}^1(X)$ *is isomorphic to the motive that corresponds to the Picard variety P_X of X under the equivalence of Theorem 2.7.2 (c). Hence* $\mathsf{ch}^1(X)$ *is unique up to isomorphism in* $\mathsf{Mot}_{rat}(k)$.

A similar result holds for the Albanese motive $\mathsf{ch}^{2d-1}(X) = (X, p_{2d-1}(X), 0)$ *and the Albanese variety A_X.*

Proof: By Remark 6.2.11 (which holds in any dimension) we may work with $p_1^{\mathrm{old}}(X)$, which we shall simply denote by $p_1(X)$. By (37) we have

$$p_1(X) = D(\beta)_C \circ i^* = p_1(X) \circ D(\beta)_C \circ (p_0(C) + p_1(C) + p_2(C)) \circ i^* \circ p_1(X).$$

Using the normalisations $D(\beta)(e) = 0$ and $^\mathsf{T}D(\beta)(e) = 0$ from Remark 6.2.5 (2), it is straightforward to check that $D(\beta)_C \circ p_0(C) = 0$ and $D(\beta)_C \circ p_2(C) = 0$. Therefore we have $p_1(X) = f \circ g$ with

$$\begin{aligned} f &= p_1(X) \circ D(\beta)_C \circ p_1(C) : \mathsf{ch}^1(C) \to \mathsf{ch}^1(X) \\ g &= p_1(C) \circ i^* \circ p_1(X) : \mathsf{ch}^1(X) \to \mathsf{ch}^1(C). \end{aligned}$$

Hence we can apply 2.3 (vii), and we obtain that $\pi = g \circ f$ is a projector, $\mathsf{ch}^1(X) \simeq (C, \pi, 0)$ and

$$\mathsf{ch}^1(C) \simeq \mathsf{ch}^1(X) \oplus (C, p_1(C) - g \circ f, 0).$$

Hence $\mathsf{ch}^1(X)$ has finite dimension in the sense of Kimura–O'Sullivan by Theorem 4.6.1 and Corollary 5.4.6.

Now turning to the relation with Theorem 2.7.2, we shall use the same notation as there. Consider $F^\natural(\pi)$. This is a projector in the category of abelian varieties up to isogeny. Using the normalizations of the divisor class β from Remark 6.2.5 (2), a straightforward calculation shows that $F^\natural(\pi) = p$ where

$$p = \frac{1}{m}\,\mathrm{pic}(i)\circ\beta\circ\mathrm{alb}(i) : J(C) \to J(C).$$

The image of this projector in the category of abelian varieties up to isogeny is, by the definition of p and by Lemma 6.2.3, isogenous with the Picard variety P_X of X. Therefore P_X corresponds to the motive $(C, \pi, 0)$ in the category M' from Theorem 2.7.2, and above we have seen that $(C, \pi, 0)$ is isomorphic to $\mathsf{ch}^1(X)$ in $\mathsf{Mot}_{\mathrm{rat}}(k)$. As the Picard variety is unique up to isomorphism by its universal property (stated in Appendix A-2), the corresponding motive in M' is unique up to isomorphism.

The assertion for $\mathsf{ch}^{2d-1}(X)$ follows by taking the transpose of the above relations.

$\square$

In the full subcategory $\mathsf{Mot}_{\mathrm{rat,fin}}(k) \subset \mathsf{Mot}_{\mathrm{rat}}(k)$ of Chow motives that are finite dimensional in the sense of Kimura–O'Sullivan we have the following result.

THEOREM 6.2.13 ([**Kimu-Mur**, Thm. 3.7]). *Let $X = X_d$ be a smooth, projective variety, and let $\mathsf{ch}^1(X) = (X, p_1(X), 0)$ be the Picard motive as constructed above. Let $N = (X, q, 0) \in \mathsf{Mot}_{\mathrm{rat,fin}}(k)$ be such that q is a lifting in $\mathsf{CH}^*(X \times X)_\mathbb{Q}$ of the first Künneth component. Then $q\circ p_1(X) : \mathsf{ch}^1(X) \to N$ and $p_1(X)\circ q : N \to \mathsf{ch}^1(X)$ are inverse isomorphisms in $\mathsf{Mot}_{\mathrm{rat}}(k)$. A similar statement holds for the Albanese motive.*

The main ingredients of the proof are
(1) the fact that $\mathsf{ch}^1(X) \in \mathsf{Mot}_{\mathrm{rat,fin}}(k)$;
(2) the results by S.–I. Kimura on motives in $\mathsf{Mot}_{\mathrm{rat,fin}}(k)$ and morphisms $f \in \mathsf{CH}^*(X \times X)_\mathbb{Q}$ that are numerically equivalent to zero;
(3) a reduction of the question whether the above morphisms in $\mathsf{Mot}_{\mathrm{rat}}(k)$ are isomorphisms to the same question in $\mathsf{Mot}_{\mathrm{hom}}(k)$ (where it is clearly true).

6.3. The Case of Surfaces

THEOREM 6.3.1 ([**Mur90**]). *Let S be a smooth, projective, irreducible surface defined over k with a rational point $e \in S(k)$. Then S has a Chow–Künneth decomposition.*

Proof: By the previous results we have projectors $p_0(S)$, $p_1(S)$, $p_3(S)$ and $p_4(S)$ orthogonal to each other. Define $p_2(S) = \Delta(S) - p_0(S) - p_1(S) - p_3(S) - p_4(S)$. Then $p_2(S)$ is a projector, orthogonal to the other ones. Clearly this gives a Chow–Künneth decomposition. $\square$

6.3.1. Distribution of the Chow Groups over the Chow–Künneth Motives. Put $\mathsf{ch}^i(S) = (S, p_i, 0)$ $(i = 0, \ldots, 4)$. By Theorem 6.2.1 (ii) and (iii) we have

$$H^i(\mathsf{ch}^j(S)) = \begin{cases} 0 & j \neq i \\ H^i(S) & j = i. \end{cases}$$

It follows from the previous results (Theorem 6.2.1) that the Chow groups $\mathsf{CH}^j(S)_\mathbb{Q}$ are distributed over the $\mathsf{ch}^i(S)$ as follows:

M	$\mathsf{ch}^0(S)$	$\mathsf{ch}^1(S)$	$\mathsf{ch}^2(S)$	$\mathsf{ch}^3(S)$	$\mathsf{ch}^4(S)$
$\mathsf{CH}^0(M)$	$\mathsf{CH}^0(S)_{\mathbb{Q}}$	0	0	0	0
$\mathsf{CH}^1(M)$	0	$\mathrm{Pic}^0_S(k)_{\mathbb{Q}}$	$\mathrm{NS}(S)_{\mathbb{Q}}$	0	0
$\mathsf{CH}^2(M)$	0	0	$T(S)(k)_{\mathbb{Q}}$	$\mathrm{Alb}_S(k)_{\mathbb{Q}}$	$\mathrm{Nm}(S)_{\mathbb{Q}}$

where as before Pic^0_S is the Picard variety, Alb_S the Albanese variety, $\mathrm{NS}(S) = \mathsf{CH}^1(S)/\mathsf{CH}^1_{\mathrm{alg}}(S)$ is the Néron–Severi group, $T(S) = \mathrm{Ker}\{\mathsf{CH}^2_{\mathrm{alg}}(S) \to \mathrm{Alb}_S\}$ is the Albanese kernel and

$$\mathrm{Nm}(S) = \mathsf{CH}^2(S)/\mathsf{CH}^2_{\mathrm{num}}(S) = \mathsf{CH}^2(S))/\mathsf{CH}^2_{\mathrm{alg}}(S).$$

Hence $\mathrm{Nm}(S)_{\mathbb{Q}} = \mathbb{Q}$; moreover the above table remains valid after a base extension $K \supset k$.

6.3.2. Refined Chow–Künneth Decomposition [Kahn-M-P].

PROPOSITION 6.3.2. *At the cost of a finite algebraic field extension (which we shall neglect in the sequel) we can split $p_2(S)$ uniquely into*

$$p_2(S) = p_2(S)^{\mathrm{alg}} + p_2(S)^{\mathrm{trans}},$$

orthogonal to each other and to the other $p_i(S)$ and such that the distribution of the Chow groups and their cohomology is as in the following table.

M	$\mathsf{ch}^2_{\mathrm{alg}}(S)$	$\mathsf{ch}^2_{\mathrm{trans}}(S)$
$H^2(M)$	$H^2_{\mathrm{alg}}(S)$	$H^2_{\mathrm{trans}}S$
$\mathsf{CH}^0(M)$	0	0
$\mathsf{CH}^1(M)$	$\mathrm{NS}(S)_{\mathbb{Q}}$	0
$\mathsf{CH}^2(M)$	0	$T(S)(k)_{\mathbb{Q}}$

This splitting leads to a unique splitting of Chow motives[7]

$$\mathsf{ch}^2(S) = \mathsf{ch}^2_{\mathrm{alg}}(S) \oplus \mathsf{ch}^2_{\mathrm{trans}}(S).$$

We have $\mathsf{ch}^2_{\mathrm{alg}}(S) \simeq \underbrace{\mathbf{L} \oplus \cdots \oplus \mathbf{L}}_{\rho\ summands}$ with ρ the Picard number of S.

Proof: Recall that $H^2(S) = H^2_{\mathrm{alg}}(S) \oplus H^2_{\mathrm{trans}}(S)$, where $H^2_{\mathrm{alg}}(S) = \mathrm{NS}(S) \otimes_{\mathbb{Q}} F$ (F is the coefficient field of the Weil cohomology theory) and $H^2_{\mathrm{trans}}(S)$ is the orthogonal complement of $H^2_{\mathrm{alg}}(S)$ in $H^2(S)$ with respect to the cup product (or equivalently, the quotient of $H^2(S)$ by $H^2_{\mathrm{alg}}(S)$). Now take an orthogonal basis of divisor classes D_i $(i = 1, \ldots, \rho)$.[8] Moreover we take the D_i such that $p_1(X)(D_i) = 0$ for all i; this is always possible, by replacing D_i–if necessary–by $D_i - p_1(X)(D_i)$. (This operation does not change its class in $\mathrm{NS}(S)_{\mathbb{Q}}$ since $p_1(X)(D_i) \in \mathsf{CH}^1_{\mathrm{alg}}(S)_{\mathbb{Q}}$.) Now take

$$p_2^{\mathrm{alg}}(S) = \sum_{i=1}^{\rho} \frac{1}{\#(D_i.D_i)} D_i \times D_i.$$

Then p_2^{alg} is a projector, orthogonal to p_0, p_1, p_3 and p_4, and we have

$$p_2^{\mathrm{alg}} {\circ} p_2 = p_2 {\circ} p_2^{\mathrm{alg}} = p_2^{\mathrm{alg}},$$

---

[7]Note that $\mathsf{ch}^2_{\mathrm{trans}}(S)$ is sometimes denoted by $t^2(S)$.

[8]It is at this stage that we possibly need to extend the original base field k.

i.e., p_2^{alg} is a constituent of p_2. Finally put $p_2^{\text{trans}} = p_2 - p_2^{\text{alg}}$. Clearly p_2^{trans} is again a projector, orthogonal to p_i ($i \neq 2$) and

$$p_2^{\text{trans}} \circ p_2 = p_2 \circ p_2^{\text{trans}} = p_2^{\text{trans}},$$

i.e., p_2^{trans} is also a constituent of p_2 and is the "complement" of p_2^{alg}. Put $\mathsf{ch}_{\text{alg}}^2(S) = (S, p_2^{\text{alg}}, 0)$ and $\mathsf{ch}_{\text{trans}}^2(S) = (S, p_2^{\text{tr}}, 0)$. Then

$$\mathsf{ch}^2(S) = \mathsf{ch}_{\text{alg}}^2(S) \oplus \mathsf{ch}_{\text{trans}}^2(S).$$

One easily checks that each summand

$$\left(S, \frac{1}{\#(D_i . D_i)} D_i \times D_i, 0 \right)$$

is isomorphic to the Lefschetz motive $\mathbf{L}$ and hence $\mathsf{ch}_{\text{alg}}^2(S)$ is a direct sum of ρ Lefschetz motives. $\square$

Remark. Note that $\mathsf{ch}_{\text{trans}}^2(S)$ is the "mysterious" part of $\mathsf{ch}^2(S)$ that is "responsible" for the *transcendental cohomology* $H_{\text{trans}}^2(S)$ and for the *Albanese kernel* $T(S)$. (And if we would have constructed motives with $\mathbb{Z}[\frac{1}{m}]$–coefficients, then also for the part of the *Brauer group* prime to m, see [**Zar**, Appendix by Mumford to Chap.VI, p. 152].)

It is immediately clear that if $\mathsf{ch}_{\text{trans}}^2(S) = 0$ then $H^2(S)_{\text{trans}} = 0$ and $T(S)(K) = 0$ for every field extension $K \supset k$ (use Remark 6.2.2 after Theorem 6.2.1). The question whether conversely $H^2(S)_{\text{trans}} = 0$ implies $\mathsf{ch}_{\text{trans}}^2(S) = 0$ is the famous *conjecture of Bloch* [**Blo80**, Lecture 1 and Appendix to Lecture 1], which is still open. One can ask the easier question

$$T(S)(K) = 0 \text{ for all fields } K \supset k \overset{?}{\Longrightarrow} \mathsf{ch}_{\text{trans}}^2(S) = 0.$$

This is indeed true, as we shall see later in Chapter 7 (see Cor. 7.6.8). For the moment we only prove that *its cohomology vanishes*.

To do this, let us first introduce the notion of *degenerate degree 0 correspondences*[9] (see [**Ful**, p. 309, Ex. 16.1.2(b)]) from X_d to Y_d as the subgroup

$$\mathsf{CH}_{\underline{\underline{\equiv}}}^d(X \times Y) \subset \mathsf{CH}^d(X \times Y) \tag{40}$$

generated by correspondences Γ for which $\mathrm{pr}_X(\Gamma) \neq X$ or $\mathrm{pr}_Y(\Gamma) \neq Y$ (or both).

Next, we prove the following auxiliary result, which goes back (although formulated in a somewhat different way) to Bloch [**Blo79**] and Bloch-Srinivas [**Blo-Sri**].

PROPOSITION 6.3.3. *Let X_d be a smooth, projective, irreducible variety defined over k. As before, let $T(X) = \mathrm{Ker}\{\mathsf{CH}_{\text{hom}}^d(X) \to \mathrm{Alb}_X\}$ denote the Albanese kernel. Assume that $T(X)(K) = 0$ for all fields $K \supset k$. Then the diagonal $\Delta(X)$ is a degenerate correspondence.*

Proof: Let ξ be the generic point of X over k, and let $e \in X(k)$ be a k–rational point (we assume for simplicity that such a point exists). Consider the cycle $(\xi) - (e) \in \mathsf{CH}^d(X_K)$, $K = k(\xi)$. Apply $p_{2d-1}(X)$ and write

$$\alpha = p_{2d-1}(X)_*((\xi) - (e)) \in \mathsf{CH}_{\text{hom}}^d(X_K).$$

---

[9]This notion is analogous to the notion of degenerate divisors on products (cf. Appendix A-2).

Note that the cycle α is supported on $C \subset X$, where C is the curve used in the construction of $p_1(X)$ and $p_{2d-1}(X)$, because $p_{2d-1}(X)$ is supported on $X \times C$. Now consider the zero–cycle

$$\zeta = (\xi) - (e) - \alpha = (\xi) - (e) - p_{2d-1}(X)((\xi) - (e)).$$

This cycle is is defined over $K = k(\xi)$ and is clearly in $\mathrm{Ker}(p_{2d-1}(X))$, hence by Theorem 6.2.1 (see also the part of the proof just above Remark 6.2.8), it also lies in the Albanese kernel $T(X)(K)$. By our assumption its class (still denoted by ζ) is zero in $\mathsf{CH}^d(X_K)$.

We now consider this cycle $\zeta = (\xi) - (e) - \alpha$ as a zero-cycle on the generic fiber $X_\xi(\simeq X_K)$ of the projection map $\mathrm{pr}_1 : X \times X \to X$, and we take its k-Zariski closure in $X \times X$ (i.e., we "spread out" the cycle ζ in $X \times X$). We obtain the cycle

$$Z = \Delta - X \times e - A$$

in $\mathsf{CH}^d(X \times X)$, with A a cycle supported on $X \times C$. Since the restriction of Z to the generic fiber is zero, and since

$$\mathsf{CH}^d(X_K) = \varinjlim_U \mathsf{CH}^d(U \times X)$$

with k-Zariski open sets $U \subset X$, we get

$$Z = \Gamma_1 \in \mathsf{CH}^d(X \times X)$$

with $\Gamma_1 \in \mathsf{CH}^d(X \times X)$ a cycle supported on $Y \times X$ for some k-Zariski closed subset $Y \subsetneq X$. Hence we have the desired result

$$\Delta = \Gamma_1 + \Gamma_2$$

with $\Gamma_2 = A + X \times e$. $\qquad\square$

Finally, we reach our aim:

COROLLARY 6.3.4. *Let X be a smooth surface. Assume that $T(X)(K) = 0$ for all fields $K \supset k$. Then $H^2_{\mathrm{trans}}(X) = 0$.*

Proof: (outline) Since Γ_1 and Γ_2 are degenerate, they act trivially on $H^2_{\mathrm{trans}}(X)$ by [**Blo80**, Appendix to Lect. 1, Lemma 1.A.5]. But the diagonal acts as the identity, hence $H^2_{\mathrm{trans}}(X) = 0$. $\qquad\square$

Appendix C: Chow-Künneth Decomposition in a Special Case

PROPOSITION. *Suppose that $X = X_d \in \mathsf{SmProj}(k)$ is a smooth projective variety such that $H^i(X)$ is algebraic for all $i \neq d$. Then X admits a self-dual Chow-Künneth decomposition. This applies in particular to smooth hypersurfaces, and, more generally, complete intersections.*

This depends on the following easy lemma, which follows directly from the definition of the composition of correspondences.

LEMMA. *Let $X = X_d \in \mathsf{SmProj}(k)$, and let $Z_i \in \mathsf{CH}^{r_i}(X)$ $(i = 1, 2, 3, 4)$ be algebraic cycles. Consider the correspondences $Z_1 \times Z_2$ and $Z_3 \times Z_4$ in $\mathsf{Corr}(X, X)$. Then*

$$(Z_3 \times Z_4)\circ(Z_1 \times Z_2) = \#(Z_2 \cdot Z_3)(Z_1 \times Z_4)$$

where $\#(Z_2.Z_3)$ denotes the intersection number.

Proof: (of the Proposition) For $i = 2p$ even and $i < d$, choose a basis e_i^ℓ ($\ell = 1, \ldots, b_i(X)$) for $H^i(X)$, with $b_i(X)$ the i-th Betti number of X, and a dual basis $\hat{e}_i^\ell$ for $H^{2d-i}(X)$. By construction we have

$$e_\ell^i \cup \hat{e}_j^m = \begin{cases} 0 & (i, \ell) \neq (j, m) \\ 1 & (i, \ell) = (j, m). \end{cases}$$

By assumption there exists algebraic cycles $E_\ell^i \in \mathsf{CH}^p(X)_{\mathbb{Q}}$ (resp. $\hat{E}_\ell^i \in \mathsf{CH}^{d-p}(X)_{\mathbb{Q}}$) such that

$$\gamma(E_\ell^i) = e_\ell^i, \quad \gamma(\hat{E}_\ell^i) = \hat{e}_\ell^i.$$

Now consider the elements

$$p_{i\ell} = E_\ell^i \times \hat{E}_\ell^i, \quad q_{i\ell} = \hat{E}_\ell^i \times E_\ell^i$$

in $\mathsf{CH}^d(X \times X)_{\mathbb{Q}}$. The above lemma shows that the $p_{i\ell}$ and $q_{i\ell}$ are mutually orthogonal projectors.

For $i = 2p$, $i < d$ put

$$p_i = \sum_\ell p_{i\ell}, \quad p_{2d-i} = \sum_\ell q_{i\ell}$$

and for i odd, $i \neq d$, put $p_i = 0$. Finally put

$$p_d = \Delta(X) - \sum_{i \neq d} p_i.$$

Then these are projectors that lift the corresponding Künneth components and that give the required Chow-Künneth decomposition. $\qquad\square$

CHAPTER 7

On the Conjectural Bloch-Beilinson Filtration

Introduction

This Chapter partly deals with *conjectures* (presented in sections 7.1 and 7.2), but also with a number of *unconditional results* related to these conjectures that will be discussed in sections 7.3, 7.4, 7.5 and 7.6.

7.1. Bloch-Beilinson Filtration

Bloch in his 1979 Duke lectures [**Blo80**] and – independently – Beilinson [**Beil**] conjectured that there exists a *descending filtration* on the Chow groups $\mathsf{CH}^i(X)_\mathbb{Q}$; this *still conjectural* filtration is nowadays called the *Bloch-Beilinson filtration*. More precisely, based upon the aforementioned paper of Beilinson on height pairings [**Beil**, 5.10]), Uwe Jannsen listed the following properties required for such a filtration [**Jann94**].

Bloch-Beilinson conjectures (version 1). Consider the category $\mathsf{SmVar}(k)$ of smooth, projective varieties defined over a field k. Given $X \in \mathsf{SmVar}(k)$ and $j \geq 0$, there exists conjecturally a descending filtration $F_{BB}^\bullet$ on $\mathsf{CH}^j(X)_\mathbb{Q}$ with the following properties:

a) $F_{BB}^0\mathsf{CH}^j(X)_\mathbb{Q} = \mathsf{CH}^j(X)_\mathbb{Q}$, $F_{BB}^1\mathsf{CH}^j(X)_\mathbb{Q} = \mathsf{CH}_{\mathrm{hom}}^j(X)_\mathbb{Q}$ (resp. $F_{BB}^1 = \mathsf{CH}_{\mathrm{num}}^j(X)_\mathbb{Q}$);

b) $F_{BB}^r\mathsf{CH}^i(X)_\mathbb{Q} \cdot F_{BB}^s\mathsf{CH}^j(X)_\mathbb{Q} \subseteq F_{BB}^{r+s}\mathsf{CH}^{i+j}(X)_\mathbb{Q}$, where we recall that "$\cdot$" denotes the intersection product of the cycle classes;

c) $F_{BB}^\bullet$ is respected by f^* and f_* for a morphism $f : X \to Y$;

d) assuming the *Künneth conjecture* from § 3.1.1 and § 6.1.1, i.e., assuming that the Künneth components Δ_i^{topo} of the diagonal $\Delta(X)$ are algebraic, we require that Δ_i operates on $\mathrm{Gr}_{BB}^\nu \mathsf{CH}^j(X)_\mathbb{Q}$ as $\delta_{i,2j-\nu}.\,\mathrm{id}$.

e) $F_{BB}^{j+1}\mathsf{CH}^j(X)_\mathbb{Q} = 0$.

Remarks.   (i) The precise meaning of part d) is as follows. By properties b) and c) the filtration $F_{BB}^\bullet$ is respected by the action of correspondences, and by a) the induced action on $\mathrm{Gr}_{BB}^\nu \mathsf{CH}^j(X)_\mathbb{Q}$ factors through homological (resp. numerical) equivalence. Part d) then means that $\mathrm{Gr}_{BB}^\nu \mathsf{CH}^j(X)_\mathbb{Q}$ depends only on the *homological motive* $h_{\mathrm{hom}}^{2j-\nu}(X) = (X, \Delta_{2j-\nu}, 0)$ (resp. on the numerical motive $h_{\mathrm{num}}^{2j-\nu}(X)$).

(ii) Part e) is sometimes replaced by the weaker requirement

e') $F_{BB}^N\mathsf{CH}^j(X)_\mathbb{Q} = 0$ for $N \gg 0$.

However Jannsen [**Jann94**, p. 258, Lemma 2.3] showed that if the standard conjecture $B(X)$ is true then e') implies e).

85

Beilinson based the above conjectures on the – still conjectural – existence of a category $\mathcal{MM}(k)$ of *mixed motives*, i.e., a category of motives for *all* varieties (not necessarily smooth, neither complete) which should be *abelian* (but *not* semisimple!).

In fact there is the following more precise version of the Bloch-Beilinson conjectures [**Jann94**, p. 259]:

Bloch-Beilinson conjectures (version 2): Conjectures a), b), c) and e) as in version 1 above, but d) is replaced by

d^+) There exists an abelian category $\mathcal{MM}(k)$ of *mixed motives* over k containing the category $\mathsf{Mot}_{\mathrm{hom}}(k)$ (resp. $\mathsf{Mot}_{num}(k)$) as a full subcategory and a functorial isomorphism

$$\mathrm{Gr}^{\nu}_{BB} \, \mathsf{CH}^j(X)_{\mathbb{Q}} \cong \mathrm{Ext}^{\nu}_{\mathcal{MM}(k)}(1, h^{2j-\nu}(X)(j)) \tag{41}$$

where $1 = (\mathrm{Spec}\, k, \mathrm{id}, 0)$ in $\mathsf{Mot}_{\mathrm{hom}}(k)$ (resp. $\mathsf{Mot}_{num}(k)$).

Formula (41) is the famous *Beilinson formula*. For its (conjectural) "consequences", see [loc. cit., sections 2 and 3].

7.2. Another Set of Conjectures

The following set of conjectures, due to the first author, stems from around 1990; it was published in 1993 in [**Mur93**] but already distributed and discussed at the Seattle conference in 1991 [**Jann94**, section 5].

In the following, X is a *smooth, projective* variety defined over a field k. For simplicty we assume that X has a k-rational point $e \in X(k)$. Furthermore we assume that we have chosen a Weil cohomology theory (for instance, $H^*(X) = H^*_{\text{ét}}(X_{\overline{k}}, \mathbb{Q}_\ell)$ with $\ell \neq \mathrm{char}(k)$). Finally we denote by K an arbitrary extension field of k.

7.2.1. Conjecture I(X) (=Conjecture A in [Mur93]).

This is the *Chow-Künneth conjecture $CK(X)$* from § 6.1.1: X has a Chow-Künneth decomposition defined over k. Recall that this means that there exist correspondences $p_i(X) \in \mathsf{Corr}^0(X, X) = \mathsf{CH}^d(X \times X)_{\mathbb{Q}}$ for $0 \leq i \leq 2d$ that are mutually orthogonal projectors, lifting the Künneth components of the diagonal and summing up to $\Delta(X) \in \mathsf{CH}^d(X \times X)_{\mathbb{Q}}$.

Assuming this, the projectors $p_i = p_i(X)$ operate on each of the Chow groups $\mathsf{CH}^j(X)_{\mathbb{Q}}$, and in fact on $\mathsf{CH}^j(X_K)_{\mathbb{Q}}$ for every extension field $K \supset k$ $(j = 0, \ldots, d)$.

7.2.2. Conjecture II(X) or "Vanishing conjecture" (=Conjecture B in [loc. cit]).

For every j $(0 \leq j \leq d)$ the projectors $p_{2d}, p_{2d-1}, \ldots, p_{2j+1}$ and $p_0, p_1, \ldots, p_{j-1}$ operate as zero on $\mathsf{CH}^j(X_K)_{\mathbb{Q}}$ for every $K \supseteq k$.

Consequences. Assuming conjectures I(X) and II(X) we have the following filtration on $\mathsf{CH}^j(X)_{\mathbb{Q}}$ (or more generally on $\mathsf{CH}^j(X_K)_{\mathbb{Q}}$):

Define

$$\begin{aligned}
F^0\mathsf{CH}^j(X)_{\mathbb{Q}} &= \mathsf{CH}^j(X)_{\mathbb{Q}} \\
F^1\mathsf{CH}^j(X)_{\mathbb{Q}} &= \mathrm{Ker}(p_{2j}) \\
F^2\mathsf{CH}^j(X)_{\mathbb{Q}} &= \mathrm{Ker}(p_{2j}) \cap \mathrm{Ker}(p_{2j-1}) \\
&\vdots \\
F^\nu\mathsf{CH}^j(X)_{\mathbb{Q}} &= \mathrm{Ker}(p_{2j}) \cap \ldots \cap \mathrm{Ker}(p_{2j-\nu+1}).
\end{aligned} \tag{42}$$

This is clearly a *descending filtration* on $\mathsf{CH}^j(X)_{\mathbb{Q}}$.

Remark. As before, put $\mathsf{ch}^i(X) = (X, p_i, 0)$. Since the projectors form a mutually orthogonal set, we have for $k \neq \ell$ that $\mathrm{Im}\, p_k \subset \mathrm{Ker}\, p_\ell$ and so, abbreviating F^* for $F^*\mathsf{CH}^j(X)_{\mathbb{Q}}$, we have (fixing j):

$$\begin{aligned}
\mathrm{Gr}_F^\nu\, \mathsf{CH}^j(X)_{\mathbb{Q}} &= \frac{F^\nu}{F^\nu \cap \mathrm{Ker}(p_{2j-\nu})} = \mathrm{Im}(p_{2j-\nu}|F^\nu) \\
&= \mathrm{Im}(p_{2j-\nu}) = \mathsf{CH}^j(\mathsf{ch}^{2j-\nu}(X)), \\
F^\nu\mathsf{CH}^j(X)_{\mathbb{Q}} &= \bigoplus_{\nu \leq \mu \leq j} \mathsf{CH}^j(\mathsf{ch}^{2j-\mu}(X)).
\end{aligned} \tag{43}$$

LEMMA 7.2.1. *Assuming conjectures $I(X)$ and $II(X)$, we have*
  (i) $F^{j+1}\mathsf{CH}^j(X)_{\mathbb{Q}} = 0$;
  (ii) $F^1\mathsf{CH}^j(X)_{\mathbb{Q}} \subseteq \mathsf{CH}^j_{\mathrm{hom}}(X)_{\mathbb{Q}}$.

Proof: Assuming II, part (i) is immediate from the definitions since $\sum_{i=0}^{2d} p_i = \Delta(X) = \mathrm{id}_X$; part (ii) follows from the commutative diagram (where γ_X denotes the cycle class map)

$$\begin{CD}
\mathsf{CH}^j(X)_{\mathbb{Q}} @>{p_{2j}}>> \mathsf{CH}^j(X)_{\mathbb{Q}} \\
@V{\gamma_X}VV @VV{\gamma_X}V \\
H^{2j}(X) @>{\Delta_{2j}^{\mathrm{topo}}}>> H^{2j}(X)
\end{CD}$$

since p_{2j} (mod homological equiv.) $= \Delta_{2j}^{\mathrm{topo}}$ operates as the identity on $H^{2j}(X)$. $\square$

7.2.3. Conjecture III(X) (=Conjecture D in [loc. cit.]) We have

$$F^1\mathsf{CH}^j(X_K)_{\mathbb{Q}} = \mathsf{CH}^j_{\mathrm{hom}}(X_K)_{\mathbb{Q}}$$

for every field extension $K \supset k$.

Remark. This seems to be the most difficult one of the conjectures, because it relates the "geometric" theory of cycles to the "cohomological" theory.

7.2.4. Conjecture IV(X) (=Conjecture C in [loc.cit.]) The filtration $F^\bullet$ defined above is independent of the ambiguity in the choice of the projectors p_i.

Remarks. (1) Conjecture IV may seem difficult, since there is – as we have seen – a lot of ambiguity in the choice of the projectors p_i. However it is based on the "philosophy" that – in spite of this ambiguity – the motives $\mathsf{ch}^i(X) = (X, p_i, 0)$ should be unique up to (natural) isomorphism; see later, Cor. 7.5.8), where we indeed show that this follows from the conjectures.

(2) In fact, if conjectures I, II and III hold for all powers of X, then IV(X) is also true (see Corollary 7.5.9).

(3) In § 7.5 we shall see that there is a close relation between the Bloch-Beilinson conjectures and the conjectures I-IV.

7.3. Some Evidence for the Conjectures; an Overview

7.3.1. Low Dimensional Varieties. Conjectures I, II, III and IV are trivially true for curves. Less trivially, they hold also for surfaces by Theorem 6.2.1. Namely by Theorem 6.3.1 surfaces have a Chow-Künneth decomposition, hence $I(S)$ is true for all surfaces S. Next let us first consider *divisor classes*: we have seen in Theorem 6.2.1 that $p_0(S)$, $p_3(S)$ and $p_4(S)$ operate as zero on $\mathsf{CH}^1(S)_\mathbb{Q}$, so $II(S)$ holds for divisors. By Lemma 7.2.1 we always have $\operatorname{Ker} p_2(S) \subseteq \mathsf{CH}^1_{\mathrm{hom}}(S)_\mathbb{Q}$. However since $\mathsf{CH}^1_{\mathrm{hom}}(S)_\mathbb{Q} = \mathsf{CH}^1_{\mathrm{alg}}(S)_\mathbb{Q}$ and since $p_1(S)$ is the identity on $\mathsf{CH}^1_{\mathrm{alg}}(S)_\mathbb{Q}$ we have for $D \in \mathsf{CH}^1_{\mathrm{hom}}(S)_\mathbb{Q}$ that

$$0 = (p_2(S){\circ}p_1(S))_*(D) = p_2(S)_*(D),$$

hence $\operatorname{Ker} p_2(S) = \mathsf{CH}^1_{\mathrm{hom}}(S)_\mathbb{Q}$ and $III(S)$ is true for divisors. Conjecture $IV(S)$ holds for divisors since the filtration on $\mathsf{CH}^1(S)_\mathbb{Q}$ is the natural one

$$\mathsf{CH}^1(S)_\mathbb{Q} \supset \mathsf{CH}^1_{\mathrm{hom}}(S)_\mathbb{Q} \supset (0).$$

Next for *zero-cycles*, we have indeed that $p_0(S)$ and $p_1(S)$ act as zero by Theorem 6.2.1, hence $II(S)$ holds. Also $III(S)$ is trivially true, and finally $IV(S)$ is true since by Theorem 6.2.1 (iii) the kernel of $p_2(S)$ on $\mathsf{CH}^2(S)_\mathbb{Q}$ is the Albanese kernel $T(S)_\mathbb{Q}$; hence the filtration on $\mathsf{CH}^2(S)_\mathbb{Q}$ is the natural one

$$\mathsf{CH}^2(S)_\mathbb{Q} \supset \mathsf{CH}^2_{\mathrm{hom}}(S)_\mathbb{Q} \supset T(S)_\mathbb{Q} \supset (0).$$

7.3.2. Special Varieties. For conjecture I, the Chow-Künneth conjecture, we refer to the list in 6.1.5. For the other conjectures we have the following **facts**:

(1) In § 7.4 we shall show that conjectures I, II and III are true for threefolds of the form $X = S \times C$, where S is a surface and C a curve (always assumed smooth and projective).

(2) For fourfolds of the form $X = S \times S'$, with S and S' surfaces, conjecture I is clearly true (see 6.1.5 (3)) but also some important consequences of conjecture II are true. See § 7.6 for details.

(3) For an abelian variety A of dimension g conjecture I is true provided one takes the "canonical projectors" from [**Den-Mu**]; concerning conjecture II we have that $p_i(A)$ operates as zero on $\mathsf{CH}^j(A)_\mathbb{Q}$ for $i < j$ and $i > j+g$. In the remaining range $2j < i < j + g$ the conjecture II coincides with a conjecture of Beauville [**Beau83**], and if this part is also true then $IV(A)$ is true.

We shall give some information on the *Proof*. For conjecture I, see § 6.1.5. For conjecture II see [**Mur93**, part I, section 2.5]. This part depends on the so-called Fourier theory of Mukai and the results of Beauville on the decomposition of the "eigenspaces" of $\mathsf{CH}^j(A)_\mathbb{Q}$ under the action of multiplication by n on A. Assuming conjecture $II(A)$ we get for $IV(A)$ the following [**Mur93**, part I, Lemma 2.5.4]: the filtration in this case is given by $F^\nu \mathsf{CH}^j(A)_\mathbb{Q} = \oplus_{s>\nu} \mathsf{CH}^j_s(A)$ $(0 \leq \nu \leq j)$, where

$$\mathsf{CH}^j_s(A) = \{Z \mid n^*(Z) = n^{2j-s}Z\},$$

and $F^{j+1} = 0$.

7.3.3. Behaviour on Divisors and Zero-cycles.

LEMMA 7.3.1. *Given* $X \in \mathsf{SmVar}(k)$, *let* $p_0(X)$ *and* $p_1(X)$ *be the projectors constructed in section* 6.2. *Assume that* $p_0(X)$ *and* $p_1(X)$ *can be completed to a full set of projectors that give a Chow-Künneth decomposition of* X. *Then conjectures II, III and IV are true for* $\mathsf{CH}^1(X)_\mathbb{Q}$.

Proof: By Theorem 6.2.1 the projector $p_1(X)$ operates as the identity on $\mathsf{CH}^1_{\mathrm{alg}}(X)_\mathbb{Q}$. Now let $i \neq 1$ and $D \in \mathsf{CH}^1_{\mathrm{alg}}(X)_\mathbb{Q}$. Then

$$0 = (p_i(X) \circ p_1(X))_*(D) = p_i(X)_*(D).$$

Hence for $i \neq 1$ the projector $p_i(X)$ acts as zero on $\mathsf{CH}^1_{\mathrm{alg}}(X)_\mathbb{Q} = \mathsf{CH}^1_{\mathrm{hom}}(X)_\mathbb{Q}$. Applying this for $i = 2$ we see that $\mathrm{Ker}\, p_2(X)$ on $\mathsf{CH}^1(X)_\mathbb{Q}$ is indeed $\mathsf{CH}^1_{\mathrm{hom}}(X)_\mathbb{Q}$, so conjecture III is true. For $i \neq 1,2$ we put $D_i = p_i(X)(D)$. Then $0 = (p_2(X) \circ p_i(X))_*(D) = p_2(X)(D_i)$, hence $D_i \in \mathsf{CH}^1_{\mathrm{hom}}(X)_\mathbb{Q} = \mathsf{CH}^1_{\mathrm{alg}}(X)_\mathbb{Q}$. Therefore, by the above, if $i \neq 1,2$ then

$$p_i(X)_*(D) = (p_i(X) \circ p_i(X))_*(D) = p_i(X)_*(D_i) = 0$$

and conjecture II is true. Finally the induced filtration on $\mathsf{CH}^1(X)_\mathbb{Q}$ is the natural one $\mathsf{CH}^1(X)_\mathbb{Q} \supset \mathsf{CH}^1_{\mathrm{hom}}(X)_\mathbb{Q} \supset (0)$, hence conjecture IV is true. $\qquad\square$

Remark 7.3.2. Let X be a smooth, projective variety of dimension d, and let $p_{2d}(X)$, $p_{2d-1}(X)$ be the projectors constructed in § 6.2. Assume that they can be completed to a full Chow-Künneth decomposition for X. What about the action of the projectors on the group of zero-cycles $\mathsf{CH}^d(X)_\mathbb{Q}$? We can not say much. Trivially $F^1\mathsf{CH}^d(X)_\mathbb{Q} = \mathsf{CH}^d_{\mathrm{hom}}(X)_\mathbb{Q}$ is the subgroup of zero-cycles of degree zero. However non-trivially we have by Theorem 6.2.1 (iii) that $F^2\mathsf{CH}^d(X)_\mathbb{Q}$ is the Albanese kernel $T(X)_\mathbb{Q}$.

7.4. Threefolds of Type $S \times C$

This section is devoted to a sketch of the proof of the following Proposition, which appeared in [**Mur93**, Part II].

PROPOSITION 7.4.1. *Let* $X = S \times C$ *with* S *a surface and* C *a curve (always tacitly assumed to be smooth and projective over* k). *Then* X *satisfies conjectures I, II and III.*

7.4.1. General Remarks on Chow-Künneth Projectors on products.
Let $Z = X_d \times Y_e$ be the product of two varieties of dimension d resp. e, hence $n = \dim Z = d + e$. Assume that X and Y have a Chow-Künneth decomposition with projectors $p_i(X)$ $(0 \leq i \leq 2d)$ and $p_j(Y)$ $(0 \leq j \leq 2e)$.

In order to distinguish the factors of the product we use the following convention:[1] write $X_1 = X_2 = X$, $Y_1 = Y_2 = Y$ and $Z_1 = Z_2 = Z$. There is the obvious isomorphism

$$X_1 \times Y_1 \times X_2 \times Y_2 \simeq X_1 \times X_2 \times Y_1 \times Y_2$$

that we shall tacitly suppress from the notation in the sequel. With this convention we have

$$Z \times Z = X_1 \times X_2 \times Y_1 \times Y_2,$$

---

[1] here the subscript does not refer to the dimension: recall that $\dim X = d$ and $\dim Y = e$

hence

$$\Delta(Z) \;=\; \Delta(X) \times \Delta(Y) = \left(\sum_{i=0}^{2d} p_i(X) \right) \times \left(\sum_{j=0}^{2e} p_j(Y) \right)$$

$$= \sum_{m=0}^{2n} \left(\sum_{i+j=m} p_i(X) \times p_j(Y) \right).$$

Hence Z has a Chow-Künneth decomposition with projectors $p_m(Z)$ $(0 \le m \le 2n)$ defined by

$$p_m(Z) = \sum_{i+j=m} p_i(X) \times p_j(Y).$$

Let us write $p_{ij}(Z) = p_i(X) \times p_j(Y)$. Then $p_m(Z) = \sum_{i+j=m} p_{ij}(Z)$.

Remark. Note that the correspondences $p_{ij}(Z)$ are themselves projectors, mutually orthogonal.

7.4.2. On the Proof of Proposition 7.4.1. A large part of the proof consists of easy, but tedious verifications. We omit most of them (see [loc. cit.] for details), but we shall prove some of the more complicated parts.

As we have seen in the previous subsection, $X = S \times C$ has a Chow-Künneth decomposition with projectors $p_m(X)$ $(0 \le m \le 6)$ that admit a description in terms of the projectors $p_i(S)$ and $p_j(C)$ from Chap. 6.

So we need to check conjectures II(X) and III(X) for the above projectors $p_m(X)$.

7.4.3. Conjectures II and III for $\mathsf{CH}^1(S \times C)_{\mathbb{Q}}$.

LEMMA 7.4.2. *If $D \in \mathsf{CH}^1_{\mathrm{hom}}(S \times C)_{\mathbb{Q}}$ then D is degenerate, i.e., $D = D_1 \times C + S \times D_2$ with $D_1 \in \mathsf{CH}^1_{\mathrm{hom}}(S)$ and $D_2 \in \mathsf{CH}^1_{\mathrm{hom}}(C)$.*

Proof: Replacing if necessary D by an integral multiple of D, we may and do assume that $D \in \mathsf{CH}^1_{\mathrm{alg}}(S \times C) \subset \mathsf{CH}^1_{\mathrm{hom}}(S \times C)$. The result then follows from Proposition A-2.2.

$\square$

PROPOSITION 7.4.3. *Conjectures II and III hold for $\mathsf{CH}^1(S \times C)_{\mathbb{Q}}$.*

Proof: Theorem 6.2.1 and Lemma 7.4.2 imply that the projector

$$p_1(X) = p_1(S) \times p_0(C) + p_0(S) \times p_1(C)$$

acts as the identity on $\mathsf{CH}^1_{\mathrm{alg}}(S \times C)_{\mathbb{Q}} = \mathsf{CH}^1_{\mathrm{hom}}(S \times C)_{\mathbb{Q}}$. Once this has been shown, the rest of the proof goes as in lemma 7.3.1; we leave this to the reader. $\square$

7.4.4. Conjectures II and III for $\mathsf{CH}^2(S \times C)_{\mathbb{Q}}$. Recall that if X, X', Y and Y' are smooth, projective varieties and $Z \in \mathsf{CH}(X \times Y) = \mathsf{Corr}(X, Y)$, $\alpha \in \mathsf{Corr}(X, X')$ and $\beta \in \mathsf{Corr}(Y, Y')$ are correspondences then

$$(\alpha \times \beta)_*(Z) = \beta \circ Z \circ {}^{\mathsf{T}}\alpha \in \mathsf{Corr}(X', Y')$$

by Lieberman's Lemma 2.1.3. In the following we shall often tacitly apply this Lemma.

Another remark: in the following proofs we shall – for convenience – sometimes follow the convention of § 7.4.1 and denote the factors by $S_1 = S_2 = S$ and $C_1 =$

$C_2 = C$ in order to distinguish them. We shall also usually consider cycles in "general position", leaving the "special positions" ("horizontal" or "vertical" cycles) to the reader.

PROPOSITION 7.4.4. *Conjecture II holds for* $\mathsf{CH}^2(S \times C)_{\mathbb{Q}}$.

Proof: We have to show that the projectors $p_0(X)$, $p_1(X)$, $p_5(X)$ and $p_6(X)$ act as zero on $\mathsf{CH}^2(S \times C)_{\mathbb{Q}}$. We shall restrict our attention to proving $p_1(X)_*(Z) = 0$ and $p_5(X)_*(Z) = 0$ for $Z \in \mathsf{CH}^2(S \times C)_{\mathbb{Q}}$; the analogous statements for $p_0(X)$ and $p_6(X)$ are easy.

Step 1. We shall first show that $p_1(X)_*(Z) = 0$ for all $Z \in \mathsf{CH}^2(S \times C)_{\mathbb{Q}}$. By 7.4.1 we have $p_1(X) = p_{10}(X) + p_{01}(X)$. We claim that both $p_{01}(X)_*(Z) = 0$ and $p_{10}(X)_*(Z) = 0$. By Lieberman's Lemma we have

$$p_{10}(X)_*(Z) = (p_1(S) \times p_0(C))_*(Z) = p_0(C) \circ Z \circ p_3(S).$$

It suffices to show that the transpose ${}^{\mathsf{T}}(p_0(C) \circ Z \circ p_3(S)) = p_1(S) \circ {}^{\mathsf{T}}Z \circ p_2(C)$ is zero. Let $e_C \in C$ be a base point. An easy calculation gives

$$
\begin{aligned}
p_1(S) \circ {}^{\mathsf{T}}Z \circ p_2(C) &= p_1(S) \circ {}^{\mathsf{T}}Z \circ (C_2 \times e_C) \\
&= p_1(S) \circ (C_2 \times {}^{\mathsf{T}}Z(e_C)) = 0 \\
&= C_2 \times p_1(S)_*({}^{\mathsf{T}}Z(e)) = 0
\end{aligned}
$$

because $p_1(S)$ acts as zero on $\mathsf{CH}^2(S)_{\mathbb{Q}}$.

Next choose a base point $e_S \in S$ and consider

$$
\begin{aligned}
p_{01}(X)_*(Z) &= (p_0(S) \times p_1(C))_*(Z) \\
&= p_1(C) \circ Z \circ p_4(S) \\
&= p_1(C) \circ Z \circ (S_2 \times e_S) \\
&= p_1(C) \circ (S_2 \times Z(e_S)) \\
&= S_2 \times p_1(C)_*(Z(e_S)) = 0
\end{aligned}
$$

because $Z(e_S) \in \mathsf{CH}^2(C)_{\mathbb{Q}} = 0$.

Step 2. We show that $p_5(X)_*(Z) = 0$. From 7.4.1 we have $p_5(X) = p_{41}(X) + p_{32}(X)$. Again we claim that both correspondences act as zero on $\mathsf{CH}^2(S \times C)_{\mathbb{Q}}$. First consider

$$
\begin{aligned}
p_{41}(X)_*(Z) &= (p_4(S) \times p_1(C))_*(Z) \\
&= p_1(C) \circ Z \circ {}^{\mathsf{T}}p_4(S) \\
&= p_1(C) \circ Z \circ p_0(S).
\end{aligned}
$$

Now consider first $Z \circ p_0(S)$. A calculation shows that $Z \circ p_0(S) = m.(e_S \times C)$, where m is the degree of the map $\mathrm{pr}_C : Z \to C$. Next $p_1(C) \circ (e_S \times C) = e_S \times p_1(C)_*(C) = 0$ since $p_1(C)_*(C) = 0$. Hence $p_{41}(X)_*(Z) = 0$.

Finally we have to show that $p_{32}(X)_*(Z) = 0$. We have

$$p_{32}(X)_*(Z) = (p_3(S) \times p_2(C))_*(Z) = p_2(C) \circ Z \circ {}^{\mathsf{T}}p_3(S) = p_2(C) \circ Z \circ p_1(S).$$

Now we first compute $p_2(C) \circ Z = (C \times e_C) \circ Z = \mathrm{pr}_S(Z) \times e_C$. Next

$$(\mathrm{pr}_S(Z) \times e_C) \circ p_1(S) = {}^{\mathsf{T}}p_1(S)_*(\mathrm{pr}_S(Z)) \times e_C = p_3(S)_*(\mathrm{pr}_S(Z)) \times e_C = 0$$

because $p_3(S)_*$ acts as zero on divisors.

Conclusion. Conjecture II is true for $\mathsf{CH}^2(S \times C)_{\mathbb{Q}}$. $\square$

LEMMA 7.4.5. *Conjecture III is true for* $\mathsf{CH}^2(S \times C)_{\mathbb{Q}}$.

Proof: We have to show: if $Z \in \mathsf{CH}^2_{\text{hom}}(S \times C)_{\mathbb{Q}}$ then $p_4(X)_*(Z) = 0$. By 7.4.1 we have

$$p_4(X) = p_{40}(X) + p_{31}(X) + p_{22}(X)$$

with $p_{ij}(X) = p_i(S) \times p_j(C)$.

(i) We first show that $p_{40}(X)_*(Z) = 0$. By Lieberman's Lemma we have

$$(p_4(S) \times p_0(C))_*(Z) = p_0(C) \circ Z \circ {}^{\mathsf{T}} p_4(S) = p_0(C) \circ Z \circ p_0(S).$$

Let $e_S \in S$ be a base point. We have $Z \circ p_0(S) = m(e_S \times C)$ with m the degree of Z over C. As Z is homologically equivalent to zero we have $m = 0$, and the result follows.

(ii). $p_{31}(X)_*(Z) = 0$ (the delicate case!)

To study the action of $p_{31}(X) = p_3(S) \times p_1(C)$, we proceed in three steps.

Step 1. Case $Z = \alpha \times C$ with α a zero-cycle on S. Then

$$(p_3(S) \times p_1(C))_*(\alpha \times C) = p_3(S)_*(\alpha) \times p_1(C)_*(C) = 0$$

because $p_1(C)_*(C) = 0$. (Remark: for the proof we do not need $\deg(\alpha) = 0$.)

Step 2. Case $Z = D \times \alpha$ with D a divisor on S, $\alpha \in C$. Then

$$(p_3(S) \times p_1(C))_*(D \times \alpha) = p_3(S)_*(D) \times p_1(C)_*(\alpha) = 0$$

because $p_3(S)_*(D) = 0$. (Remark: again the proof works for arbitrary D and α.)

Step 3. General case.

Put

$$Z' = p_{31}(X)_*(Z) = p_1(C) \circ Z \circ p_1(S).$$

We need to show that $Z' = 0$ if Z is homologically equivalent to zero. Let $Z_1 = {}^{\mathsf{T}} Z' \in \mathsf{Corr}^0(C, S)$. For every point $x \in C$ we have $\deg(Z_1(x)) = 0$ since Z_1, being the transpose of Z, is homologically equivalent to zero. Moreover, using Step 1 we may assume without loss of generality that $Z_1(e_C) = 0$ for a point e_C on C.

Claim. For $\eta \in C$ generic point we have $Z_1(\eta) = 0$.

Proof of Claim: Z_1 determines a homomorphism of abelian varieties

$$\varphi_{Z_1} : J(C) \to \mathrm{Alb}(S).$$

We claim first that $\varphi_{Z_1} = 0$. It suffices to check this statement on the torsion points of $J(C)$. We have

$$\varphi_{Z_1} : J(C)[\ell^n] = H^1_{\text{ét}}(C_{\bar{k}}, \mathbb{Z}/\ell^n\mathbb{Z}) \to \mathrm{Alb}_S[\ell^n] = H^3_{\text{ét}}(S_{\bar{k}}, \mathbb{Z}/\ell^n\mathbb{Z})$$

which is given by the Künneth component $\gamma(Z_1)_{13}$ of the cycle class $\gamma(Z_1)$. Since Z_1 is homologically equivalent to zero we have $\varphi_{Z_1} = 0$, and hence $Z_1(\eta)$ is albanese equivalent to zero on S. But $Z_1 = {}^{\mathsf{T}} Z' = p_3(S) \circ {}^{\mathsf{T}} Z \circ p_1(C)$, hence $Z_1(\eta) = p_3(S)_* {}^{\mathsf{T}} Z(\eta - e_C)$ and since $p_3(S)$ is a projector we have $Z_1(\eta) = p_3(S)_* Z_1(\eta)$. Since $Z_1(\eta)$ is albanese equivalent to zero on S and the albanese kernel is also the kernel of $p_3(S)_*$ by Theorem 6.2.1 (iii) we get $Z_1(\eta) = 0$, which proves the claim.

Since $\mathsf{CH}^2(S_\eta)_{\mathbb{Q}} = \varinjlim_U \mathsf{CH}^2(U \times S)_{\mathbb{Q}}$ where U runs through the open sets of C, we get $Z_1 = \sum_i a_i \times D_i$ with $a_i \in C$ and D_i divisors on S. Hence $Z' = {}^{\mathsf{T}} Z_1 = \sum_i D_i \times a_i$. Finally note that $Z' = (p_{31}(X))_*(Z)$ and since $p_{31}(X)$ is a projector we have $Z' = p_{31}(X))_*(\sum_i D_i \times a_i)$. Hence $Z' = 0$ by Step 2.

Remark. Note that in the above proof we have used the inclusion $T(S)_\mathbb{Q} \subset \operatorname{Ker} p_3(S)$, where $T(S)$ denotes the albanese kernel. This is the "hard" part of the proof of Theorem 6.2.1.(iii).

(iii) $p_{22}(X)_*(Z) = 0$. We have

$$p_{22}(X)_*(Z) = p_2(C) \circ Z \circ p_2(S).$$

Consider first $p_2(C) \circ Z = \operatorname{pr}_S(Z) \times e_C$ and note that $\operatorname{pr}_S Z$ is a divisor on S homologically equivalent to zero. We finally get

$$p_2(C) \circ Z \circ p_2(S) = p_2(S)_*(\operatorname{pr}_S(Z)) \times e_C = 0$$

because $\operatorname{Ker} p_2(S) = \mathsf{CH}^1_{\mathrm{hom}}(S)_\mathbb{Q}$.

This completes the proof of Lemma 7.4.4 and of Proposition 7.4.1 as far as $\mathsf{CH}^2(S \times C)_\mathbb{Q}$ is concerned. $\qquad\square$

7.4.5. Conjectures II and III are true for $\mathsf{CH}^3(S \times C)_\mathbb{Q}$. This is easy and is left to the reader (see [**Mur93**, part II, section 8.4]).

7.5. On Some Results of Jannsen and Their Consequences

In 1991 Uwe Jannsen proved the following theorem.

THEOREM 7.5.1 ([**Jann94**, Thm. 5.2, p.288]). *Fix a field k. Then the Bloch-Beilinson conjectures a),b),c),d) and e) from § 7.1 are true for all smooth, projective varieties if and only if this is so for the conjectures I, II, III and IV from § 7.2. Moreover, if these conjectures are true, then the Bloch-Beilinson filtration is the same as the filtration from § 7.2; in fact, a filtration with these properties is unique ([ibid., Cor. 5.7]).*

For the proof we refer to [ibid], section 5. In the proof a crucial rôle is played by the following proposition.

PROPOSITION 7.5.2 ([**Jann94**, Proposition 5.8]). *Let X_d and Y_e be smooth, projective varieties that have a Chow-Künneth decomposition given by*

$$\mathsf{ch}^i(X) = (X, p_i(X), 0)(0 \le i \le 2d) \ resp. \ \mathsf{ch}^j(Y) = (Y, p_j(Y), 0)(0 \le j \le 2e).$$

Consider the product variety $Z = X \times Y$, which has then a Chow-Künneth decomposition with projectors

$$p_m(Z) = \sum_{r+s=m} p_r(X) \times p_s(Y), \quad 0 \le m \le 2d + 2e$$

as we have seen in § 7.4.1. Now the following holds:

 a) *If, with these projectors, Conjecture II(Z) holds then*

$$\operatorname{Hom}_{\mathsf{Mot}_{\mathrm{rat}}(k)}(\mathsf{ch}^i(X), \mathsf{ch}^j(Y)) = 0 \ \ \text{for } i < j.$$

 b) *If, with these projectors, Conjecture III(Z) holds then*

$$\operatorname{Hom}_{\mathsf{Mot}_{\mathrm{rat}}(k)}(\mathsf{ch}^i(X), \mathsf{ch}^i(Y)) \xrightarrow{\sim} \operatorname{Hom}_{\mathsf{Mot}_{\mathrm{hom}}(k)}(h^i(X), h^i(Y)).$$

Remarks. 1) In fact in Part a) we only need that $p_m(Z)$ acts as zero on $\mathsf{CH}^d(Z)_{\mathbb{Q}}$ for $m > 2d$.

Part b) of the Proposition is equivalent to the following statement: if $R \in \mathsf{CH}^d(X \times Y)_{\mathbb{Q}} = \mathsf{Corr}^0(X, Y)$ is such that $p_i(Y) \circ R \circ p_i(X)$ is homologically equivalent to zero, then $p_i(Y) \circ R \circ p_i(X) = 0$. This holds in particular if R itself is homologically equivalent to zero.

2) In matrix form this Proposition can be expressed as follows. Given $R \in \mathsf{CH}^d(X \times Y)$, write

$$\alpha_{ji}(R) = p_j(Y) \circ R \circ p_i(X) : \mathsf{ch}^i(X) \to \mathsf{ch}^j(Y).$$

Then if Z satisifies conjectures II and III the matrix has the following form (for simplicity we first take $d = e$, so that we have a square matrix)

$$(\alpha_{ji}(R)) = \begin{pmatrix} r = h & * & * & \dots & * \\ 0 & r = h & * & \dots & * \\ \vdots & & \ddots & \ddots & \ddots & \vdots \\ \vdots & & & \ddots & \ddots & * \\ 0 & \dots & \dots & 0 & r = h \end{pmatrix},$$

i.e., in the lower left hand corner $(j > i)$ we have zeroes and on the diagonal $(j = i)$ the correspondence $\alpha_{ii}(R)$ depends only on its class modulo homological equivalence.

In the general case (d and e arbitrary) we have a similar pattern, but now we have a matrix with $2e + 1$ rows and $2d + 1$ columns.

Proof of Proposition 7.5.2. For the proof in the general case we refer to [**Jann94**, p. 293-294]. Here we give the proof under the extra assumption that the Chow-Künneth decomposition of X is *self-dual*, i.e., $p_{2d-i}(X) = {}^{\mathsf{T}}p_i(X)$ (which is for us the most important case!).

For simplicity of notation, write

$$M_i = \mathsf{ch}^i(X), \ N_j = \mathsf{ch}^j(Y), \ p_i = p_i(X) \text{ and } p'_j = p_j(Y).$$

Now recall (see 7.4.1) that, with the convention that we suppress the obvious isomorphism $X \times Y \times X \times Y \cong (X \times X) \times (Y \times Y)$, we have

$$\Delta_Z = \Delta_X \times \Delta_Y = \left(\sum_i p_i\right) \times \left(\sum_j p'_j\right) = \sum_m p_m(Z), \quad p_m(Z) := \sum_{i+j=m} p_i \times p'_j.$$

Now in $\mathsf{Mot}_{\mathrm{rat}}(k)$ we have

$$\mathrm{Hom}(M_i, N_j) = \left\{ p'_j \circ R \circ p_i \mid R \in \mathsf{CH}^d(X \times Y)_{\mathbb{Q}} \right\}.$$

The action of $p_m(Z) = \sum_{r+s=m} p_r \times p'_s$ on such a correspondence is given by

$$p_m(Z)_*(p'_j \circ R \circ p_i) = \sum_{r+s=m} (p_r \times p'_s)_*(p'_j \circ R \circ p_i).$$

By Lieberman's Lemma (see Lemma 2.1.3) we get

$$\sum_{r+s=m} (p_r \times p'_s)_*(p'_j \circ R \circ p_i) = \sum_{r+s=m} p'_s \circ p'_j \circ R \circ p_i \circ {}^{\mathsf{T}}p_r = p'_j \circ R \circ p_i$$

because in the sum only the term with $s = j$ and $2d - r = i$ survives due to the extra assumption ${}^{\mathsf{T}}p_r = p_{2d-r}$ (and the others give zero). Now $r + s = m$, hence

$2d - i + j = m$. Therefore if $\mathrm{II}(Z)$ holds we get zero if $m > 2d$, i.e., if $i < j$, because the homomorphism $p_m(Z)$ acts on $\mathsf{CH}^d(Z)_{\mathbb{Q}}$.

For the same reason we get that if $\mathrm{III}(Z)$ holds and if R is homologically equivalent to zero then

$$p_{2d}(Z)_*(p_i' \circ R \circ p_i) = p_i' \circ R \circ p_i = 0;$$

in fact, we only need that $p_i' \circ R \circ p_i$ is homologically equivalent to zero. $\square$

COROLLARY 7.5.3. *Let* $X = X_d$ *(always smooth and projective). Assume that conjectures I, II and III hold for* X *and* $X \times X$. *Let* $f \in \mathsf{Corr}^0_{\mathsf{Mot}_{\mathrm{rat}}(k)}(X, X)$. *If* f *is homologically equivalent to zero then* f *is nilpotent, in fact* $f^{2d+1} = 0$.

Proof: With the above notation the matrix for f is a $(2d + 1) \times (2d + 1)$-matrix of the form

$$(\alpha_{ji}(f)) = \begin{pmatrix} 0 & * & \cdots & * \\ & 0 & & \vdots \\ \vdots & & \ddots & * \\ 0 & \cdots & 0 & 0 \end{pmatrix}.$$

Hence $f^{2d+1} = 0$. $\square$

COROLLARY 7.5.4. *Assume that conjectures I, II and III hold for* $X = X_d$ *and* $X \times X$. *Consider the two-sided ideal* $I(X)$ *defined by the exact sequence*

$$0 \to I(X) \to \mathsf{Corr}^0_{\mathsf{Mot}_{\mathrm{rat}}(k)}(X, X) \to \mathsf{Corr}^0_{\mathsf{Mot}_{\mathrm{hom}}(k)}(X, X) \to 0.$$

Then $I(X)$ *is nilpotent, in fact* $I(X)^{2d+1} = 0$.

Proof: This follows from the previous Corollary. $\square$

COROLLARY 7.5.5 (No phantom motives exist!). *Assume that* $X \times X$ *satisfies I, II and III. Let* $M = (X, p, 0) \in \mathsf{Mot}_{\mathrm{rat}}(k)$ *be a Chow motive such that its image in* $\mathsf{Mot}_{\mathrm{hom}}(k)$ *is zero. Then* $M = 0$.

Proof: By assumption p is homologically equivalent to zero. Hence p is nilpotent by Corollary 7.5.3. But since p is a projector it follows that $p = 0$. $\square$

COROLLARY 7.5.6. *Assume that* $X \times X$ *satisfies I, II and III. Let* $M = (X, p, 0) \in \mathsf{Mot}_{\mathrm{rat}}(k)$. *If* $H(M) = 0$ *(i.e., M has no cohomology) then* $M = 0$.

Proof: Consider the cycle class $\gamma_{X \times X}(p) \in H^{2d}(X \times X)$. Since $H(M) = 0$ all the Künneth components of $\gamma_{X \times X}(p)$ are zero. Hence $\gamma_{X \times X}(p) = 0$, i.e., p is homologically equivalent to zero. Hence $p = 0$ by the previous corollary. $\square$

COROLLARY 7.5.7. *Assume that conjectures I, II and III hold for* X *and all powers of* X. *Then the motives of type* $(X, p, n) \in \mathsf{Mot}_{\mathrm{rat}}(k)$ *are finite dimensional. In particular, these conjectures imply the Kimura-O'Sullivan Conjecture (Conjecture 5.6.8).*

Proof: Because of Thm. 5.4.4 it suffices to do this for $M = \mathsf{ch}(X) = (X, \mathrm{id}, 0)$. Since X satisfies the the C-K conjecture, it satisfies in particular the sign conjecture § 4.5. So we can split $M = M_+ \oplus M_-$ with where M_+ and M_- has only even, respectively odd cohomology. It follows Cor. 4.3.4 that for sufficiently large n we have $H(\bigwedge^n M_+) = H(\mathsf{Sym}^n(M_-)) = 0$ and so, by Cor. 7.5.6, $\bigwedge^n M = \mathsf{Sym}^n(M) = 0$ and hence M is finite dimensional. $\square$

COROLLARY 7.5.8. *Assume that conjectures I, II and III hold for X and all powers of X. Let $M = (X, p, 0)$ and $M' = (X, p', 0)$ be two Chow motives such that $p = p'$ modulo homological equivalence. Then $f = p' \circ p : M \to M'$ and $g = p \circ p' : M' \to M$ are natural isomorphisms in $\mathsf{Mot}_{\mathrm{rat}}(k)$. This holds in particular for the Chow-Künneth projectors p_i and p'_i (for all i).*

Proof: From I, II and III it folllows that M and M' are finite dimensional by Corollary 7.5.7. The result then follows from Theorem A in [**Kimu-Mur**], which states that if $f : M \to N$ and $g : N \to M$ are morphisms in $\mathsf{Mot}_{\mathrm{rat}}(k)$ between finite dimensional motives and f_{hom} and g_{hom} are isomorphisms in $\mathsf{Mot}_{\mathrm{hom}}(k)$, then f and g are isomorphisms in $\mathsf{Mot}_{\mathrm{rat}}(k)$. $\qquad\square$

COROLLARY 7.5.9. *Assume that conjectures I, II and III hold for X and all powers of X. Then the filtration is unique, i.e., conjecture $IV(X)$ holds.*

Proof: Suppose we have two C-K decompositions $\mathsf{ch}(X) = \oplus(X, p_i, 0) = \oplus(X, p'_i, 0)$ with two corresponding filtrations F^* and $(F')^*$. By symmetry it suffices to show that $F'^{\nu} \subset F^{\nu}$. Apply Proposition 7.5.2 a) with $Y = X$, $\mathsf{ch}(X) = \oplus(X, p_i, 0)$ and $\mathsf{ch}(Y) = \oplus(X, p'_i, 0)$. It gives

$$p_{2j-t} \circ p'_{2j-s} = p_{2j-t} \circ \mathrm{id}_X \circ p'_{2j-s} = 0 \quad \text{for } s \geq \nu \text{ and } t < \nu,$$

hence on $\mathsf{CH}^j(X)_{\mathbb{Q}}$ we have the inclusion

$$\mathrm{Im}(p'_{2j-s}) \subset \mathrm{Ker}(p_{2j-t}).$$

for all $s \geq \nu$ and $t < \nu$ and

$$(F')^{\nu} = \bigoplus_{s \geq \nu} \mathrm{Im}(p_{2j-s}) \subseteq \cap_{t < \nu} \mathrm{Ker}(p_{2j-t}) = F^{\nu};$$

cf. equations (42) and (43). $\qquad\square$

COROLLARY 7.5.10 (Bloch's conjecture). *Let S be a smooth, projective surface such that $H^2(S)_{\mathrm{trans}} = 0$. If conjecture III holds for $S \times S$ then the albanese kernel (with rational coefficients) $T(S)_{\mathbb{Q}}$ is zero; in fact, $\mathsf{ch}^2_{\mathrm{trans}}(S) = 0$.*

Proof: We have seen in 6.3 that S has a Chow-Künneth decomposition. If $H^2(S)_{\mathrm{trans}} = 0$ then the projector $p_2^{\mathrm{trans}}(S)$ from § 6.3.2 is homologically equivalent to zero. This means that $\mathsf{ch}^2_{\mathrm{trans}}(S)$ is a phantom motive. We want to apply Corollary 7.5.5. So we need the validity of conjectures II and III for $S \times S$. Conjecture $II(S \times S)$ indeed holds as we shall see below (c.f. Proposition 7.6.1) where it is shown for the relevant part of the Chow group $\mathsf{CH}^2(S \times S)_{\mathbb{Q}}$). Hence if also conjecture $III(S \times S)$ holds then $p_2^{\mathrm{trans}}(S) = 0$ by Corollary 7.5.5. $\qquad\square$

7.6. Products of Two Surfaces

In this section S and S' are smooth, projective irreducible surfaces defined over a field k. Furthermore $C = C(S)$ (resp. $C' = C(S')$) is the hyperplane section of S (resp. of S') used in the construction of the projectors $p_1(S)$ and $p_3(S)$ of S (resp. $p_1(S')$ and $p_3(S')$ (see § 6.2).

Remarks.   (1) Note that we have replaced p_j^{old} by p_j^{new}, $j = 1, 2$. So, to be correct, if C occurs in connection with p_1 and p_3 respectively, by what we did in § 6.2 (just after Lemma 6.2.10 where we treat the case of surfaces) we should replace C by

$$C_1^{\text{new}} := C \cup \bigcup_i {}^{\mathsf{T}}D(\beta)(a_i), \text{ resp. } C_3^{\text{new}} := C \cup \bigcup_i D(\beta)(a_i), \ C \cdot C = \sum_i a_i.$$

So, in what follows we should take the support on $C_1^{\text{new}} \times S$, respectively $S \times C_3^{\text{new}}$ instead of $C \times S$, respectively $S \times C$. For simplicity we have ignored below such fineries. This also holds for the curve C' on S'.

(2) In some of the following proofs we shall tacitly use Fulton's theory of refined intersections, which allows us to control the support of the intersections.

We consider the variety $X = S \times S'$ with Chow-Künneth projectors

$$p_m(X) = \sum_{i+j=m} p_i(S) \times p_j(S').$$

In order to simplify the notation we often write $p_i = p_i(S)$ and $p'_j = p_j(S')$. Let $R \in \mathsf{CH}^2(S \times S')_{\mathbb{Q}} = \mathsf{Corr}^0(S, S')$ and write

$$\alpha_{ji}(R) = p'_j {\circ} R {\circ} p_i \ \ (i, j = 0, 1, 2, 3, 4)$$

as in § 7.5.

PROPOSITION 7.6.1. ([**Kahn-M-P**, Thm. 7.3.10] *and also* [**Kim**] *With the above assumptions and notations, conjecture II is true for* $\mathsf{CH}^2(S \times S')_{\mathbb{Q}}$.

COROLLARY 7.6.2. *We have*

$$\mathrm{Hom}_{\mathsf{Mot}_{\mathrm{rat}}(k)}(\mathsf{ch}^i(S), \mathsf{ch}^j(S')) = 0$$

for $i < j$.

Proof of the above Corollary: This follows immediately from Propositions 7.6.1 and 7.5.2 a. $\qquad\qquad\square$

Remarks. (1) With the notations from the remark following Proposition 7.5.2 it follows that for $R \in \mathsf{CH}^2(S \times S')_{\mathbb{Q}}$ the 5 by 5 matrix $(\alpha_{ji}(R)$ has the following shape.

$$(\alpha_{ji}(R)) = \begin{pmatrix} * & * & * & 0 & 0 \\ 0 & * & * & * & 0 \\ 0 & 0 & * & * & * \\ 0 & 0 & 0 & * & * \\ 0 & 0 & 0 & 0 & * \end{pmatrix}.$$

Note that we have also three zeroes in the upper right hand corner by the same argument as in the proof of Prop. 7.5.2 since the projectors $p_0(S \times S')$ and $p_1(S \times S')$ act also as zero on $\mathsf{CH}^2(S \times S')_{\mathbb{Q}}$.

(2) In fact $\alpha_{00}(R), \alpha_{11}(R), \ \alpha_{33}(R)$ and $\alpha_{44}(R)$ depend only on homological equivalence. Compare [**Kahn-M-P**, p. 163].

The following trivial lemma will be used in the proof of Proposition 7.6.1.

LEMMA 7.6.3. *Let* $A \in \mathsf{CH}(S)_{\mathbb{Q}}$, $B \in \mathsf{CH}(S')_{\mathbb{Q}}$ *be cycles, and let* $\alpha \in \mathsf{Corr}(S, S)$, $\beta \in \mathsf{Corr}(S', S')$ *be correspondences. We have*

$$(\alpha \times \beta)_*(A \times B) = \alpha_*(A) \times \beta_*(B).$$

Proof: (of Proposition 7.6.1) We shall only prove that $p_5(X)$ and $p_1(X)$ act as zero on $\mathsf{CH}^2(X)_{\mathbb{Q}}$; the cases $p_0(X)$ and $p_6(X)$, $p_7(X)$, $p_8(X)$ are easier and are left to the reader.

Step 1. The correspondence $p_5(X)$ acts as zero on $\mathsf{CH}^2(X)_{\mathbb{Q}}$.

We have $p_5(X) = p_{41}(X) + p_{32}(X) + p_{23}(X) + p_{14}(X)$. Let $R \in \mathsf{CH}^2(S \times S')_{\mathbb{Q}}$.

Case A) p_{41} and p_{14} act as zero.

Proof: We first consider p_{41}. Let $e_S \in S$ be a base point. We have

$$p_{41}(X)_*(R) = p_1' \circ R \circ {}^{\mathsf{T}}p_4 = p_1' \circ R \circ p_0 = p_1' \circ R \circ (e_S \times S') = p_1' \circ (e_S \times \mathrm{pr}_{S'}(R)).$$

Suppose that R is nondegenerate, i.e., $\mathrm{pr}_{S'}(R) = nS'$. Apply Lemma 7.6.3 with $\alpha = \mathrm{id}_S$, $\beta = p_1'$, $A = e_S$ and $B = \mathrm{pr}_{S'}(R) = n.[S']$. Then we get

$$p_1' \circ (e_S \times \mathrm{pr}_{S'})(R) = p_1' \circ (e_S \times nS') = e_S \times np_1'(S') = e_S \times 0 = 0.$$

We leave it to the reader to check the assertion when R is degenerate, i.e. if $R \in \mathsf{CH}^2(S \times S')_{\mathbb{Q}}$ is of type $R_1 \times R_1'$ with R_1 (resp. R_1') a divisor on S (resp. S').

Next consider p_{14}. We have $p_{14}(X)_*(R) = p_4' \circ R \circ p_1$. Take the transpose to obtain

$$^{\mathsf{T}}p_{14}(X)_*(R) = {}^{\mathsf{T}}p_1 \circ {}^{\mathsf{T}}R \circ {}^{\mathsf{T}}p_4' = p_3 \circ {}^{\mathsf{T}}R \circ p_0'.$$

As in the case of p_{41} one shows that $^{\mathsf{T}}p_{14}(X)_*(R) = 0$ using $p_3(S)(S) = 0$. $\square$

Case B). Next consider the (more serious) cases $p_{32}(X)$ and $p_{23}(X)$. We claim that both act as zero.

Proof: Consider

$$p_{32}(X)_*(R) = p_2' \circ R \circ {}^{\mathsf{T}}p_3 = p_2' \circ R \circ p_1.$$

First of all recall equation (39):

$$p_1(S) = p_1^{\mathrm{new}}(S) = p_1^{\mathrm{old}}(S) - \tfrac{1}{2}p_1^{\mathrm{old}}(S)p_3^{\mathrm{old}}(S).$$

It suffices to prove $p_2' \circ R \circ p_1^{\mathrm{old}} = 0$ because then also $p_2' \circ R \circ p_1^{\mathrm{old}} \circ p_3^{\mathrm{old}} = 0$. Therefore we shall drop the subscript "old" and write $p_1 = p_1^{\mathrm{old}}(S)$. For p_1 we use the construction given in § 6.2; in particular we shall use the divisor $D(\beta)_C \in \mathsf{CH}^1(C \times S)_{\mathbb{Q}}$ introduced in formula (37). Recall that $p_1 = D(\beta)_C \circ i_C^*$, where $i_C : C \to S$ is the inclusion of the hyperplane section. We get $p_2' \circ R \circ p_1 = p_2' \circ R \circ D(\beta)_C \circ i_C^*$. Consider first the divisor

$$D = p_2' \circ R \circ D(\beta)_C \in \mathsf{CH}^1(C \times S')_{\mathbb{Q}}$$

and the corresponding morphism $D : \mathsf{ch}(C) \to \mathsf{ch}^2(S')$. Now we shall show that $D = 0$, which clearly implies that $p_2' \circ R \circ p_1 = p_{32}(X)_*(R) = 0$. The morphism $D : \mathsf{ch}(C) \to \mathsf{ch}^2(S')$ determines correspondences $D_r = D \circ p_r(C)$ $(r = 0, 1, 2)$ and morphisms (denoted by the same letter)

$$D_r : \mathsf{ch}^r(C) \to \mathsf{ch}^2(S') \quad (r = 0, 1, 2).$$

In Proposition 7.4.4 we have seen that the variety $Y = C \times S' \cong S' \times C$ satisfies conjecture II for all Chow groups, in particular for divisors. So it follows from Proposition 7.5.2 (Jannsen's proposition) that $D_r = 0$ for $r = 0, 1$. Hence

$$D = D_2 = D \circ p_2(C) = D \circ (C \times e_C) = p_2' \circ R \circ D(\beta)_C \circ (C \times e_C).$$

Now consider $D(\beta)_C \circ (C \times e_C) = C \times D(\beta)_C(e_C)$. As $D(\beta)_C(e_C) = 0$ by the normalisation in § 6.2, we have $D = 0$. For $p_{23}(X)_*(R)$ we use the transpose. This completes the proof for $p_5(X)$. $\square$

Step 2. Next consider the action of $p_1(X)$ on $\mathsf{CH}^2(S \times S')_\mathbb{Q}$. We have $p_1(X) = p_{10}(X) + p_{01}(X)$ with $p_{ij}(X) = p_i \times p'_j$. Consider

$$
\begin{aligned}
p_{01}(X)_*(R) &= p'_1 \circ R \circ {}^\mathsf{T}p_0 = p'_1 \circ R \circ p_4 = p'_1 \circ R \circ (S \times e_S) \\
&= p'_1 \circ (S \times R(e_S)) = S \times (p'_1)_*(R(e_S))
\end{aligned}
$$

and note that this is zero because $R(e_S) \in \mathsf{CH}^2(S')_\mathbb{Q}$ and p'_1 operates as zero on zero-cycles.

Working with the transpose we also get $p_{10}(X)_*(R) = 0$. This completes the proof for $\mathsf{CH}^2(S \times S')_\mathbb{Q}$ (leaving the actions of $p_0(X)$, $p_6(X)$, $p_7(X)$ and $p_8(X)$ to the reader!) $\qquad\square$

The remaining case $\alpha_{22}(R)$. Again $R \in \mathsf{CH}^2(S \times S)_\mathbb{Q}$. What can we say about $\alpha_{22}(R) = p'_2 \circ R \circ p_2 : \mathsf{ch}^2(S) \to \mathsf{ch}^2(S')$?

LEMMA 7.6.4. *We have*

   (i) $\mathrm{Hom}_{\mathsf{Mot}_{\mathrm{rat}}(k)}(\mathsf{ch}^2_{\mathrm{alg}}(S), \mathsf{ch}^2_{\mathrm{trans}}(S')) = 0$;

   (ii) $\mathrm{Hom}_{\mathsf{Mot}_{\mathrm{rat}}(k)}(\mathsf{ch}^2_{\mathrm{trans}}(S), \mathsf{ch}^2_{\mathrm{alg}}(S')) = 0$.

Proof: For the proof of part (i), recall from § 6.3.2 that

$$
p^{\mathrm{alg}}_2(S) = \sum_{i=1}^{\rho} \frac{1}{\#(D_i \cdot D_i)} D_i \times D_i
$$

where $\{D_i\}_{i=1}^{\rho}$ is an orthogonal basis for $\mathrm{NS}(S)_\mathbb{Q}$ and $\#(D_i.D_i)$ is the self-intersection number of D_i. So let us define the motive

$$
M_i = (S, \frac{1}{\#(D_i \cdot D_i)} D_i \times D_i, 0)
$$

and consider $\mathrm{Hom}_{\mathsf{Mot}_{\mathrm{rat}}(k)}(M_i, \mathsf{ch}^2_{\mathrm{trans}}(S'))$. An element in this group is (up to scalar) of the form

$$
(p'_2)^{\mathrm{trans}} \circ R \circ (D \times D) = (p'_2)_*^{\mathrm{trans}}(D \times R_*(D))
$$

where D is a divisor on S', but this is zero since

$$
(p'_2)_*^{\mathrm{trans}}(D \times R_*(D)) = D \times (p'_2)_*^{\mathrm{trans}}(R_*(D)) = D \times 0;
$$

see § 6.3.2.

For part (ii), take the transpose to reduce to the previous case. $\qquad\square$

LEMMA 7.6.5. *We have*

$$
\mathrm{Hom}_{\mathsf{Mot}_{\mathrm{rat}}(k)}(\mathsf{ch}^2_{\mathrm{alg}}(S), \mathsf{ch}^2_{\mathrm{alg}}(S')) \xrightarrow{\sim} \mathrm{Hom}_{\mathsf{Mot}_{\mathrm{hom}}(k)}(h^2_{\mathrm{hom}}(S), h^2_{\mathrm{hom}}(S')).
$$

Proof: Let $\{D_i\}$ (resp. $\{D'_j\}$) be an orthogonal basis for $\mathrm{NS}(S)_\mathbb{Q}$ (resp. $\mathrm{NS}(S')_\mathbb{Q}$). Consider

$$
(p'_2)^{\mathrm{alg}} \circ R \circ p^{\mathrm{alg}}_2 = \sum_{i,j} \frac{1}{\#(D'_j \cdot D'_j)}(D'_j \times D'_j) \circ R \frac{1}{\#(D_i \cdot D_i)} \circ (D_i \times D_i).
$$

We must show that this is zero if R is homologically equivalent to zero. Working term by term we get

$$
\begin{aligned}
(D' \times D') \circ R \circ (D \times D) &= (D' \times D') \circ (D' \cdot R_*(D)) \\
&= \#(D' \cdot R_*(D)) \circ (D \times D')
\end{aligned}
$$

and $\#(D' \cdot R_*(D)) = 0$ if R is homologically equivalent to zero. $\qquad\square$

So there remains the crucial case

$$\mathrm{Hom}_{\mathrm{Mot}_{\mathrm{rat}}(k)}(\mathsf{ch}^2_{\mathrm{trans}}(S), \mathsf{ch}^2_{\mathrm{trans}}(S')).$$

Its elements are of the form $p_2^{\mathrm{trans}}(S')\circ R\circ p_2^{\mathrm{trans}}(S)$ with $R \in \mathsf{CH}^2(S \times S')_{\mathbb{Q}} = \mathsf{Corr}^0(S, S')$. We are not yet able to prove conjecture $\mathrm{III}(S \times S')$. However we can say something about this group. Namely, consider the group $\mathsf{CH}^2_{\underline{\underline{\equiv}}}(S \times S')_{\mathbb{Q}}$ of degenerate correspondences defined in (40), Chapter 6. Then we have the following

THEOREM 7.6.6 ([**Kahn-M-P**, Theorem 7.4.3]). *The map*

$$\alpha : \mathsf{CH}^2(S \times S')_{\mathbb{Q}} \to \mathrm{Hom}_{\mathrm{Mot}_{\mathrm{rat}}(k)}(\mathsf{ch}^2_{\mathrm{trans}}(S), \mathsf{ch}^2_{\mathrm{trans}}(S'))$$

defined by $\alpha(R) = p_2^{\mathrm{trans}}(S')\circ R\circ p_2^{\mathrm{trans}}(S)$ *induces an isomorphism*

$$\beta : \mathsf{CH}^2(S \times S')_{\mathbb{Q}}/\mathsf{CH}^2_{\underline{\underline{\equiv}}}(S \times S')_{\mathbb{Q}} \xrightarrow{\sim} \mathrm{Hom}_{\mathrm{Mot}_{\mathrm{rat}}(k)}(\mathsf{ch}^2_{\mathrm{trans}}(S), \mathsf{ch}^2_{\mathrm{trans}}(S')).$$

This theorem should be compared with the theorem of Weil A-2.1 where we have similar results for curves but where it is possible to replace the "abstract" motives $\mathsf{ch}^1(C)$ and $\mathsf{ch}^1(C')$ by the "concrete geometric" jacobians $J(C)$ and $J(C')$.

Proof of the Theorem: By definition the map α is surjective. It remains to show that α is zero on $\mathsf{CH}^2_{\underline{\underline{\equiv}}}(S \times S')_{\mathbb{Q}}$ and that the induced map β is injective.

Step 1. The map α is zero on $\mathsf{CH}^2_{\underline{\underline{\equiv}}}(S \times S')_{\mathbb{Q}}$.

Let R be in $\mathsf{CH}^2_{\underline{\underline{\equiv}}}(S \times S')_{\mathbb{Q}}$ and consider $p_2^{\mathrm{trans}}(S')\circ R\circ p_2^{\mathrm{trans}}(S)$. Consider first the case that R is supported on $Y \times S'$ with $Y \subset S$ a curve; there is no loss of generality if we assume that Y is irreducible, and moreover, taking if necessary the desingularization of Y, we may assume that Y is smooth. Let $j : Y \hookrightarrow S$ be the natural morphism. Then

$$R = (j \times \mathrm{id}_{S'})_*R_1 = R_1\circ j^*, \qquad R_1 \in \mathsf{CH}^1(Y \times S')_{\mathbb{Q}}.$$

Hence we obtain

$$\begin{aligned} p_2^{\mathrm{trans}}(S')\circ R &= p_2^{\mathrm{trans}}(S')\circ p_2(S')\circ R \\ &= p_2^{\mathrm{trans}}(S')\circ p_2(S')\circ R_1\circ j^*. \end{aligned}$$

Consider $p_2(S')\circ R_1 \in \mathsf{Corr}(Y, S')$. Let $p_i(Y)$ be the projectors that give the Chow-Künneth decomposition of Y. Since $Y \times S'$ satisfies conjecture II, by Propositions 7.4.4 and 7.5.2 we have

$$p_2(S')\circ R_1 = p_2(S')\circ R_1\circ(\sum_{i=0}^{2} p_i(Y)) = p_2(S')\circ R_1\circ p_2(Y).$$

Now $R_1\circ p_2(Y) = R_1\circ(Y\times e_Y)$ with e_Y the base point of Y, and we get $R_1\circ(Y\times e_Y) = Y \times D$ with $D := (R_1)_*(e_Y)$ a divisor on S'. Combining all this we get

$$\begin{aligned} p_2^{\mathrm{trans}}(S')\circ R &= p_2^{\mathrm{trans}}(S') \circ R_1\circ(Y \times e_Y)\circ j^* \\ &= p_2^{\mathrm{trans}}(S')\circ(Y \times D)\circ j^* \\ &= \big(Y \times p_2^{\mathrm{trans}}(S')_*D\big) \circ j^* = 0 \end{aligned}$$

since $p_2^{\mathrm{trans}}(S')_*D = 0$, D being a divisor on S'; See the table in Proposition 6.3.2.

Next consider the case where R is supported on $S \times Y'$ with Y' a curve on S'. We want to show that $p_2^{\mathrm{trans}}(S')\circ R\circ p_2^{\mathrm{trans}}(S) = 0$. Clearly it suffices to see that the transpose

$${}^{\mathsf{T}}(p_2^{\mathrm{trans}}(S')\circ R\circ p_2^{\mathrm{trans}}(S)) = p_2^{\mathrm{trans}}(S)\circ {}^{\mathsf{T}}R\circ p_2^{\mathrm{trans}}(S')$$

is zero, but this follows from the previous case.

Step 2. The map β is injective.

Suppose that $p_2^{\mathrm{trans}}(S')\circ R\circ p_2^{\mathrm{trans}}(S)=0$. Let ξ be the generic point of S. We have

$$0=(p_2^{\mathrm{trans}}(S')\circ R\circ p_2^{\mathrm{trans}}(S))_*(\xi)=p_2^{\mathrm{trans}}(S')_*[R_*\{p_2^{\mathrm{trans}}(S)_*(\xi)\}]. \qquad (44)$$

To study this expression we proceed step by step. We first consider $p_2^{\mathrm{trans}}(S)_*(\xi)$. Recall that

$$p_2^{\mathrm{trans}}(S)=\Delta(S)-e_S\times S-p_1(S)-p_3(S)-S\times e_S-\sum_i\frac{1}{\#(D_i\cdot D_i)}(D_i\times D_i)$$

where $\{D_i\}$ is an orthogonal basis for $\mathrm{NS}(S)_{\mathbb{Q}}$. As the projectors $p_0(S)$, $p_1(S)$ and $p_2^{\mathrm{alg}}(S)$ act as zero on $\mathsf{CH}^2(S)_{\mathbb{Q}}$ we get

$$p_2^{\mathrm{trans}}(S)_*(\xi)=\xi-p_3(S)_*(\xi)-e_S$$

where $p_3(S)_*(\xi)$ is a zero-cycle supported on the curve $C\subset S$ that was used in the construction of $p_1(S)$ and $p_3(S)$ (see § 6.2). Moreover, the degree of $p_3(S)_*(\xi)$ is zero (because $D(\beta)_*(e_C)=0$).

Next we consider the operation of R on $p_2^{\mathrm{trans}}(S)_*(\xi)$. We obtain

$$R_*(p_2^{\mathrm{trans}}(S)_*(\xi))=R_*(\xi)-R_*(p_3(S)_*(\xi))-R_*(e_S).$$

Note that $R_*(p_3(S)_*(\xi))$ is a zero cycle that is supported on a curve

$$\Gamma'=R_*(C)$$

on S' and Γ' is defined over k. We shall denote this zero-cycle by $Z(\xi)$.

Finally we consider the action of $p_2^{\mathrm{trans}}(S')$ on the zero-cycle $R_*(p_2(S)_*(\xi))$. Arguing as above, we get

$$p_2^{\mathrm{trans}}(S')_*(R_*(p_2^{\mathrm{trans}}(S)_*(\xi)))=R_1'(\xi)-R_2'(\xi)-R_3'(\xi)$$

where

$$\begin{aligned}
R_1'(\xi) &= R_*(\xi)-Z(\xi)-R_*(e_S)\\
R_2'(\xi) &= p_3(S')_*(R_*(\xi)-Z(\xi)-R_*(e_S))\\
R_3'(\xi) &= \deg(R_*(\xi)-Z(\xi)-R_*(e_S))e_{S'}.
\end{aligned}$$

Note that $R_2'(\xi)$ is a zero-cycle that is supported on the curve $C'\subset S'$ that was used in the construction of $p_3(S')$ and $R_3'(\xi)=0$ because $\deg R_*(\xi)=\deg R_*(e_S)$ and $\deg Z(\xi)=0$. So $R_1'(\xi)-R_2'(\xi)$ is a zero-cycle on S_ξ',and by the relation (44) we have $R_1'(\xi)-R_2'(\xi)=0$ in $\mathsf{CH}^2(S_\xi')_{\mathbb{Q}}$. Rewriting (44) we get

$$R_*(\xi)=Z(\xi)+R_*(e_S)+R_2'(\xi), \qquad (45)$$

where the right hand side is a cycle class in $\mathsf{CH}^2(S_\xi')_{\mathbb{Q}}$ supported on $\Gamma'\cup C'\subset S'$, with Γ' and C' curves defined over the ground field k. (Also note that the finite set of points $|R_*(e_S)|$ is supported on Γ'.) Now consider S_ξ' as the fiber over ξ of $S\times S'\to S$. We have

$$\mathsf{CH}^2(S_\xi')_{\mathbb{Q}}=\varinjlim_{U}\mathsf{CH}^2(U\times S')_{\mathbb{Q}}$$

where U runs through the open sets of S. By taking the Zariski closure in $S\times S'$ (i.e., we "spread out" these cycles) we obtain the relation

$$R=Z_1+Z_2$$

where Z_1, the Zariski closure of the right hand side of (45), is supported on $S \times Y'$ with $Y' \subset S'$ a curve, and Z_2 is supported on $Y \times S'$ with $Y \subset S$ a curve. Hence R belongs to the subgroup $\mathsf{CH}^2_{\equiv}(S \times S')_\mathbb{Q} \subset \mathsf{CH}^2(S \times S')_\mathbb{Q}$. $\qquad\square$

COROLLARY 7.6.7. *Let again S and S' be smooth projective surfaces defined over k. Let $k(S)$ be the function field of S and let $T(S')$ be the Albanese kernel of S'. Then there is a canonical surjection*

$$\lambda : T(S')_\mathbb{Q}(k(S)) \twoheadrightarrow \mathrm{Hom}_{\mathsf{Mot}_{\mathrm{rat}}(k)}(\mathsf{ch}^2_{\mathrm{trans}}(S), \mathsf{ch}^2_{\mathrm{trans}}(S')).$$

Proof: Let ξ be the generic point of S (hence $k(S) = k(\xi)$.) Let $\zeta \in T(S')_\mathbb{Q}(k(\xi))$ be a cycle class on S' in the albanese kernel, defined over $k(\xi)$, i.e. "rational over $k(\xi)$" in the technical sense. Consider the fibre S'_ξ over ξ in $S \times S'$ and choose a representing cycle Z_ξ for ζ. This is a 0-cycle (of degree 0) on S'_ξ defined over $k(\xi)$. Take its Zariski closure in $S \times S'$ and let $R \in \mathsf{CH}^2(S \times S')_\mathbb{Q}$ be its class. Obviously, R is not uniquely determined by ζ, but rather by its class modulo $\mathsf{CH}^2_{\equiv}(S \times S')_\mathbb{Q}$ since another representative Z'_ξ for ζ differs from R by a cycle supported on $D \times S'$, where D is a divisor on S.

Using the notation of Theorem 7.6.6, set $\lambda(\zeta) = \beta(R)$. This now is well-defined; hence, starting with the cycle class ζ from above we have defined a morphism $\beta : \mathsf{ch}^2_{\mathrm{trans}}(S) \to \mathsf{ch}^2_{\mathrm{trans}}(S')$.

It is also surjective as we show as follows. Start with

$$\rho \in \mathrm{Hom}_{\mathsf{Mot}_{\mathrm{rat}}(k)}(\mathsf{ch}^2_{\mathrm{trans}}(S), \mathsf{ch}^2_{\mathrm{trans}}(S')).$$

Then

$$\rho = p_2^{\mathrm{trans}}(S') \circ R_1 \circ p_2^{\mathrm{trans}}(S) = R'_1$$

where R_1 and R'_1 are elements of $\mathsf{CH}^2(S \times S')_\mathbb{Q}$. Take $\zeta = R'_1(\xi)$ and apply the above construction. With the above notation, we then get

$$\lambda(\zeta) = \alpha(R_1) = \beta(R'_1) = \rho.$$

Remark. We can prove even more [**Kahn-M-P**, Thm. 7.4.8]: the kernel of λ is the subgroup of $T(S')_\mathbb{Q}(k(S))$ generated by those elements in the Albanese kernel of S' which are defined over fields L/k contained in $k(S)$ with transcendence degree < 2.

COROLLARY 7.6.8. *Let S be a smooth projective surface over k. If $T(S)_\mathbb{Q}(K) = 0$ for all $K \supset k$, then $\mathsf{ch}^2_{\mathrm{trans}}(S) = 0$.*

Proof: By Cor. 7.6.7, it follows that in fact $\mathsf{ch}^2_{\mathrm{trans}}(S) = 0$ as soon as $T(S)_\mathbb{Q}(k(S)) = 0$, namely take the identity morphism of $\mathsf{ch}^2_{\mathrm{trans}}(C)$ in the right hand side of Cor. 7.6.7 with $S' = S$. This identity morphism comes from some $\tau \in T(S)_\mathbb{Q}(k(S))$, hence $\mathrm{id}_{\mathsf{ch}^2_{\mathrm{trans}}(S)} = 0$. $\qquad\square$

Relations Between the Conjectures

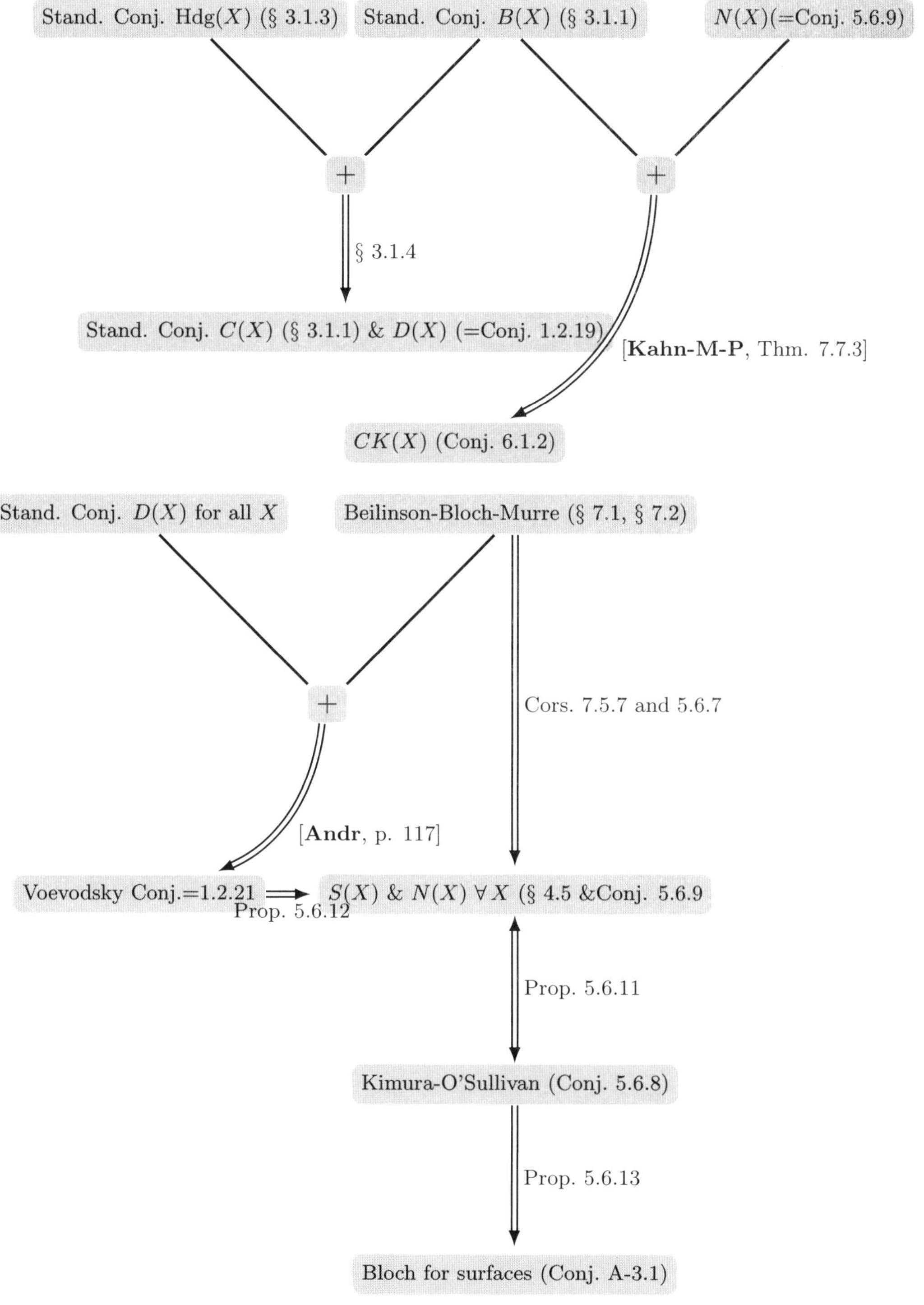

CHAPTER 8

Relative Chow-Künneth Decomposition

In this chapter we work over the complex numbers, i.e. our field $k = \mathbb{C}$. For simplicity we take for our Weil cohomology H_B^*, the Betti cohomlogy. Throughout this chapter, we write d_X for the dimension of X (assumed equi-dimensional).

8.1. Relative Motives

8.1.1. Introduction. As we have seen in Chapter 6, fixing a Weil-cohomology theory, a Chow-Künneth decomposition of $X \in \mathsf{SmProj}(k)$ is given by mutually orthogonal projectors $p_i(X) \in \mathsf{CH}(X \times X)$ with $\sum_i p_i(X) = \Delta(X)$ that give the decomposition

$$H^*(X) = \bigoplus_i H^i(X) \tag{46}$$

induced by the Künneth decomposition of the class of the diagonal $\Delta \subset X \times X$.

In the relative setting one replaces the projective variety X by a projective *morphism* $f : X \to S$ where one assumes that X is a smooth but not necessarily projective variety and S can be any variety over the field $\mathbb{C}$. We then should replace the cohomology groups $H^k(X)$ of X by the direct image sheaves $R^k f_* \mathbb{Q}$. This can be best explained using the language of derived categories as follows. For details the reader may consult [**Pe-St**, Appendix A].

—Start from any additive category $\mathfrak{A}$. In the homotopy category $\mathbf{H}(\mathfrak{A})$ the objects are the complexes of objects in $\mathfrak{A}$ while the morphisms are the equivalence classes $[f]$ of morphisms between complexes up to homotopy.[1] Suppose moreover that $\mathfrak{A}$ is abelian. Then its derived category $D(\mathfrak{A})$ is obtained from $\mathbf{H}(\mathfrak{A})$ by making quasi-isomorphisms invertible. Recall that a *quasi-isomorphism* is a morphism $s : K^\bullet \to L^\bullet$ between complexes such that the induced morphisms $H^q(s) : H^q(K^\bullet) \to H^q(L^\bullet)$ are all isomorphisms.

— Consider the category $\mathbb{Q}_X$ of sheaves of $\mathbb{Q}$-vector spaces over a topological space X. Associated to a continuous map $f : X \to S$ there is the derived direct image functor $Rf_* : D(\mathbb{Q}_X) \to D(\mathbb{Q}_S)$. Recall that for any sheaf F of $\mathbb{Q}$-vector spaces on X the object $Rf_*F \in D(\mathbb{Q}_S)$ is represented by a complex $f_*K^\bullet$ of sheaves of $\mathbb{Q}$-vector spaces on S, where f_* is the usual direct image and $K^\bullet$ is some f-acyclic resolution of F, i.e., the cohomology sheaves $\mathcal{H}^i(K^\bullet)$ vanish for all $i \geq 1$, $\mathcal{H}^0(K^\bullet) = F$, and moreover $R^i f_* K^j = 0$ for $i \geq 1$. These resolutions exist: one may take the Godement resolution of the sheaf F (see e.g [**Pe-St**, Appendix B.2]). Its cohomology sheaves are the usual direct images, e.g.

$$R^i f_* F = \mathcal{H}^i(Rf_* F) = \mathcal{H}^i(f_* K^\bullet). \tag{47}$$

Note that the direct image sheaves generalize the usual cohomology groups: if $a_X : X \to \mathrm{Spec}\, k$ is the structure morphism, then $a_{X,*} = \Gamma$ is the global section

---

[1] This notion if more fully explained in the next Chapter, see e.g. § 9.2.1.

105

functor and

$$R^k(a_X)_* \mathcal{F} = H^k(X, \mathcal{F}).$$

If $f : X \to S$ is a morphism, the Grothendieck spectral sequence associated to the commutative diagram

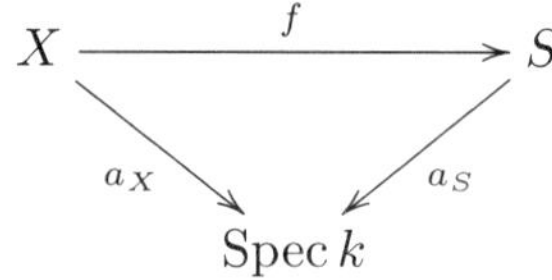

is the *Leray spectral sequence*

$$E_2^{p,q} = H^p(S, R^q f_* \mathcal{F}) \Rightarrow H^{p+q}(X, \mathcal{F}).$$

Next, one needs to generalize the decomposition (46) to the relative setting. This is fairly easy if one considers *smooth* projective morphisms: by [**Del68**] the Leray spectral sequence for f degenerates at E_2 and gives the decomposition

$$Rf_* \mathbb{Q}_X \simeq \bigoplus_i R^i f_* \mathbb{Q}_X[-i], \tag{48}$$

where the (non-canonical) isomorphism takes place in the *derived* category $D(\mathbb{Q}_S)$ of complexes of sheaves of $\mathbb{Q}$–vector spaces over S. In other words, the complex of sheaves $Rf_* \mathbb{Q}_X$ on the left can be replaced by the complex on the right with $R^i f_* \mathbb{Q}_X$ put on place i and where the differentials are all zero.

It is in this setting that Deninger and Murre [**Den-Mu**] generalized the notion of Chow-Künneth decomposition by replacing the projectors $p_i(X)$ by relative projectors $p_i(X/S) \in \mathsf{CH}_{d_X}(X \times_S X)$ that induce the above decomposition.

If one no longer assumes that f is smooth, the Leray-spectral sequence need not degenerate at E_2. However, ordinary cohomology on a *smooth* variety such as X is the same as the so-called intersection cohomology for the constant sheaf. Intersection cohomology is known to behave much better under morphisms if one is prepared to work with *perverse sheaves*. These objects in general are not sheaves at all but complexes of sheaves satisfying certain axioms which are recalled in § 8.2. Since X is smooth, the constant sheaf on X becomes perverse when viewed as a complex concentrated in degree $-d$, where $d = \dim(X)$. There is a perverse variant of the Leray spectral sequence which degenerates at E_2: this is the content of the decomposition theorem of Beilinson-Bernstein-Deligne-Gabber which we recall below (Theorem 8.2.6). Corti and Hanamura [**Cor-Ha00**] extended the work of Deninger-Murre to this setting. In their approach, a relative Chow-Künneth decomposition is given by a set of mutually orthogonal relative projectors that induce this generalized decomposition of $Rf_* \mathbb{Q}_X$.

8.1.2. Relative Correspondences.

DEFINITION 8.1.1. Let S be a quasi-projective variety over k. The category $\mathsf{Var}(S)$ is the category whose objects are pairs (X, f) with X a smooth quasi-projective variety over k and $f : X \to S$ a projective morphism. A morphism from (X, f) to (Y, g) is a morphism $h : X \to Y$ such that $g \circ h = f$.
By abuse of notation, we shall sometimes suppress the structure morphism to S from the notation and write $X \in \mathsf{Var}(S)$.

In order to generalize the notion of a degree d correspondence, recall that

$$\mathsf{Corr}^p(X, Y) = \mathsf{CH}^{p+d_X}(X \times Y)_{\mathbb{Q}} = \mathsf{CH}_{-p+d_Y}(X \times Y)_{\mathbb{Q}}.$$

The latter expression also makes sense for singular varieties adopting the definition for Chow groups from [**Ful**, Ch1.3]. This then leads to the following definition in the relative situation.

DEFINITION 8.1.2. Let (X, f), $(Y, g) \in \mathsf{Var}(S)$. Suppose that Y is equidimensional of dimension d_Y. The group

$$\mathsf{Corr}^p_S(X, Y) = \mathsf{CH}_{d_Y - p}(X \times_S Y)_{\mathbb{Q}}$$

is called the group of degree p relative correspondences (over S) from X to Y. If Y is not equidimensional, write $Y = \coprod Y_i$ with Y_i equidimensional of dimension d_i and put

$$\mathsf{Corr}^p_S(X, Y) = \bigoplus_i \mathsf{CH}_{d_i - p}(X \times_S Y_i)_{\mathbb{Q}}.$$

Remark 8.1.3. Note if $s \in S$ is a point such that the fibers X_s and Y_s over s are smooth and if $T \in \mathsf{Corr}^p_S(X, Y)$, then

$$T|_{X_s \times Y_s} \in \mathsf{Corr}^p(X_s, Y_s).$$

To define composition of relative correspondences we use Fulton's refined Gysin homomorphisms [**Ful**, § 6.2]. These are defined for cartesian squares

$$\begin{array}{ccc} Y & \longrightarrow & X \\ \downarrow & & \downarrow \\ S & \overset{i}{\longrightarrow} & T, \end{array}$$

where i is a regular embedding. If i is of codimension d, the upper map induces $i^! :$ $\mathsf{CH}_k(X)_{\mathbb{Q}} \to \mathsf{CH}_{k-d}(Y)_{\mathbb{Q}}$. We apply this construction using the following diagram with right hand side Cartesian square

$$\begin{array}{ccccc} X \times_S Z & \overset{p_{XZ}}{\longleftarrow} & X \times_S Y \times_S Z & \longrightarrow & (X \times_S Y) \times_k (Y \times_S Z) \\ & & \downarrow & & \downarrow \\ & & Y & \overset{\delta}{\longrightarrow} & Y \times_k Y. \end{array}$$

Here, since Y is smooth, δ is a regular embedding and so the refined Gysin homomorphism $\delta^!$ is well-defined. Now, given two relative correspondences $\Gamma_1 \in \mathsf{Corr}^p_S(X, Y)$ and $\Gamma_2 \in \mathsf{Corr}^q_S(Y, Z)$ we define

$$\Gamma_2 \circ_S \Gamma_1 := (p_{XZ})_*(\delta^!(\Gamma_1 \times_k \Gamma_2)) \in \mathsf{Corr}^{p+q}_S(X, Z). \tag{49}$$

8.1.3. Definition and Basic Properties.

DEFINITION 8.1.4. The objects in the category $\mathsf{Mot}_{\mathrm{rat}}(S)$ of Chow motives over S are triples $(X, p(X/S), n)$ with $X \in \mathsf{Var}(S)$, $p(X/S) \in \mathsf{Corr}^0_S(X, X)$ a projector (i.e., $p(X/S) \circ p(X/S) = p(X/S)$) and $n \in \mathbb{Z}$. Morphisms are given by

$$\mathrm{Hom}((X, p(X/S), m), (Y, q(Y/S), n) = q(Y/S) \circ \mathsf{Corr}^{n-m}_S(X, Y) \circ p(X/S).$$

Note that if $S = k$, the category $\mathsf{Mot}_{\mathrm{rat}}(\mathrm{Spec}\, k) = \mathsf{Mot}_{\mathrm{rat}}(k)$ is the category of Chow motives that we defined in Chapter 2.

EXAMPLE 8.1.5. The motive $\mathbf{L}_S = (S, \mathrm{id}_S, -1)$ is called the *Lefschetz motive* over S.

Let $\Delta_S : X \hookrightarrow X \times_S X$ be the diagonal embedding. There is a contravariant functor

$$\mathsf{ch}_S : \mathsf{Var}(S) \to \mathsf{Mot}_{\mathrm{rat}}(S)$$

that sends $X \in \mathsf{Var}(S)$ to its relative motive $\mathsf{ch}_S(X) = (X, \Delta_S(X), 0)$ and a morphism $h : X \to Y$ to its transposed graph ${}^{\mathrm{T}}\Gamma_h$.

Observe that the Hom-groups in the category of relative motives are given by

$$\mathrm{Hom}_{\mathsf{Mot}_{\mathrm{rat}}(S)}(\mathsf{ch}_S(X) \otimes \mathbf{L}_S^{\otimes i}, \mathsf{ch}_S(Y) \otimes \mathbf{L}_S^{\otimes j}) \;=\; \mathsf{Corr}_S^{i-j}(X,Y) \qquad (50)$$
$$=\; \mathsf{CH}_{d_Y - i + j}(X \times_S Y)_{\mathbb{Q}}.$$

In particular

$$\mathrm{Hom}_{\mathsf{Mot}_{\mathrm{rat}}(S)}(\mathbf{L}_S^{\otimes i}, \mathsf{ch}_S(Y)) = \mathsf{CH}_{d_Y - i}(Y)_{\mathbb{Q}}. \qquad (51)$$

Functoriality. Given a flat morphism $f : S \to S'$ there is an induced (contravariant) morphism f^* on the level of Chow groups, and for proper f there is a (covariant) morphism f_*. See [**Ful**] for details. These induced morphisms can be used to pull back or push forward relative correspondences.

LEMMA 8.1.6 (Composition of relative correspondences and base change). *Let $X, Y, Z \in \mathsf{Var}(S)$ be quasi-projective and smooth over k, and let $f : S \to S'$ be a morphism. Let $j_{XY} : X \times_S Y \hookrightarrow X \times_{S'} Y$ be the natural morphism and similarly for j_{YZ} and j_{XZ}.*

(1) *Suppose f is flat. Let $\Gamma_1' \in \mathsf{Corr}_{S'}(X,Y)$, $\Gamma_2' \in \mathsf{Corr}_{S'}(Y,Z)$. Then*

$$((j_{YZ})^*\Gamma_2') \circ_S ((j_{XY})^*\Gamma_1') = (j_{XZ})^*(\Gamma_2' \circ_{S'} \Gamma_1').$$

(2) *Suppose f is proper. Let $\Gamma_1 \in \mathsf{Corr}_S(X,Y)$, $\Gamma_2 \in \mathsf{Corr}_S(Y,Z)$. Then*

$$((j_{YZ})_*\Gamma_2) \circ_{S'} ((j_{XY})_*\Gamma_1) = (j_{XZ})_*(\Gamma_2 \circ_S \Gamma_1).$$

Proof: We prove only (2) since it is similar but more complicated than the proof of (1). Moreover, we shall only use (2).

Consider the following diagram

$$
\begin{array}{ccccc}
X \times_S Z & \xleftarrow{p_{XZ}} & X \times_S Y \times_S Z & \to & (X \times_S Y) \times_k (Y \times_S Z) \\
\downarrow{\scriptstyle j_{XZ}} & & \downarrow{\scriptstyle q} & & \downarrow{\scriptstyle p} \\
X \times_{S'} Z & \xleftarrow{p'_{XZ}} & X \times_{S'} Y \times_{S'} Z & \to & (X \times_{S'} Y) \times_k (Y \times_{S'} Z) \\
& & \downarrow & & \downarrow \\
& & Y & \xrightarrow{\delta} & Y \times_k Y.
\end{array}
$$

By [**Ful**, Thm. 6.2, p. 98] (with δ playing the rôle of i) we have

$$\delta^!(j_{XY} \times_k j_{YZ})_*(\Gamma_1 \times_k \Gamma_2) = q_*(\delta^!(\Gamma_1 \times_k \Gamma_2)).$$

Apply $(p'_{XZ})_*$ to this formula to obtain:

$$
\begin{aligned}
(j_{XZ})_*\Gamma_2 \circ_{S'} (j_{YZ})_*\Gamma_1 &= (p'_{XZ})_*(q_*(\delta^!(\Gamma_1 \times_k \Gamma_2)) \\
&= (j_{XZ})_*(p_{XZ})_*(\delta^!(\Gamma_1 \times_k \Gamma_2)) \\
&= (j_{XZ})_*(\Gamma_2 \circ_S \Gamma_1). \qquad \square
\end{aligned}
$$

COROLLARY 8.1.7. *There exists a functor $\mathsf{Mot}_{\mathrm{rat}}(S) \to \mathsf{Mot}_{\mathrm{rat}}(k)$ from the category of relative Chow motives to the category of absolute Chow motives. This* *functor sends a motive $M = (X, p(X/S), n)$ to $(X, p(X), n)$, where $p(X)$ is the image of $p(X/S)$ under the map*

$$\mathsf{CH}_{d_X}(X \times_S X)_{\mathbb{Q}} \quad \rightarrow \quad \mathsf{CH}_{d_X}(X \times X)_{\mathbb{Q}}$$
$$\| \qquad\qquad\qquad \|$$
$$\mathsf{Corr}^0_S(X, X) \qquad\qquad \mathsf{Corr}^0(X, X),$$

which is a ring homomorphism so that $p(X)$ is indeed a projector.

Proof: Apply Lemma 8.1.6 (2) to the structure morphism $a_S : S \to \operatorname{Spec} k$. $\square$

The category $\mathsf{Mot}_{\mathrm{rat}}(S)$ is an additive tensor category with sum and tensor product given by

$$\mathsf{ch}_S(X) + \mathsf{ch}_S(Y) = \mathsf{ch}_S(X \sqcup Y), \quad \mathsf{ch}_S(X) \otimes \mathsf{ch}_S(Y) = \mathsf{ch}_S(X \times_S Y).$$

8.2. Perverse Sheaves and the Decomposition Theorem

As mentioned in the introduction of this Chapter, if we work on possibly singular varieties we better replace ordinary cohomology by intersection cohomology. The definition of the latter concept is quite involved and we shall only sketch it; for details, we shall refer to [**Pe-St**].

The basic idea is as follows. If S is an irreducible singular variety, its singularities form a proper subvariety S_{sing} and hence this is a variety of lower dimension. The complement $S - S_{\mathrm{sing}}$ figures in this way as an open stratum of highest dimension in a stratification of S. The next lower dimensional strata are then constructed in the same manner on the components of the singular variety S_{sing}. The resulting stratification of S consists of only smooth strata. However, a normal slice to a fixed stratum (in the classical topology) may have non-constant topological type. It is a non-trivial theorem that one may refine the given stratification in such a way that along a fixed stratum the normal slice stays locally the same. These stratifications have further technical properties and are called *Whitney stratifications*. See [**Pe-St**, Appendix C] for a technical summary and further references.

Using such a stratification, one defines the *intersection complex* $\mathrm{IC}^{\bullet}_S$ of S [**Pe-St**, Def. 13.12]. This complex belongs to the subcategory $D^b_c(\mathbb{Q}_S)$ of the derived category of complexes of sheaves of $\mathbb{Q}$–vector spaces consisting of complexes of sheaves whose cohomology sheaves are *constructible* (i.e., locally constant on *some* stratification of S) and *bounded* (i.e., zero outside a finite range). In the case of the intersection complex, the range is $[-d_S, -1]$. The hypercohomology groups of $\mathrm{IC}^{\bullet}_S$ are called the *intersection cohomology groups* of S.

The intersection complex satisfies further properties, the support and co-support condition. The co-support condition is dual to the support condition under the so-called Verdier duality, a generalization of the (topological) Poincaré duality [**Pe-St**, §13.1]. It is an involutive operator $D_S : D_c(\mathbb{Q}_S) \to D_c(\mathbb{Q}_S)$.

DEFINITION 8.2.1. A complex $F^{\bullet} \in D^b_c(\mathbb{Q}_S)$ satisfies the *support condition* if

$$\dim \operatorname{Supp} \mathcal{H}^{-i}(F^{\bullet}) \leq i$$

for all $i \in \mathbb{Z}$. We say that $F^{\bullet}$ satisfies the *cosupport condition* if its Verdier dual $D_S(F^{\bullet})$ satisfies the support condition.

DEFINITION 8.2.2. A *perverse sheaf* is a complex $F^{\bullet} \in D^b_c(\mathbb{Q}_S)$ that satisfies both the support and cosupport conditions. The subcategory of perverse sheaves is denoted by $\mathrm{Perv}(S)$.

EXAMPLE 8.2.3. If S is smooth, every local system $\mathbb{L}$ can be viewed as a perverse sheaf by considering it as a complex concentrated in degree $-d_S$. We denote this as $\mathbb{L}[d_S]$.[2] Indeed, $\mathbb{L}[d_S]$ has cohomology only in degree $-d_S$ and its Verdier dual, $\mathbb{L}^\vee[d_S]$ only has cohomology in degree $-d_S$. So both support and co-support conditions hold since these cohomology sheaves are supported on all of S. In general, if $\mathbb{L}$ is a local system on a Zariski open dense subset U of S consisting of smooth points, there is a unique extension of $\mathbb{L}$ to a perverse sheaf, the *intersection complex* $\mathrm{IC}_S(\mathbb{L})$. In particular, if $S = U$ is smooth then $\mathrm{IC}_S(\mathbb{L}) = \mathbb{L}[d_S]$.

Unfortunately, the usual cohomology sheaves of a perverse complex need not be perverse at all. To remedy this, one needs *perverse* cohomology sheaves. Below we give a brief outline of their construction; for details see [**Beil-Ber-Del**].

Recall that the usual cohomology of a complex K in an abelian category $\mathfrak{A}$ can be obtained using the classical *truncation functors* $\tau_{\leq k}$ and $\tau_{\geq k}$. These are obtained as follows: for $\tau_{\leq k}$, remove the K^p with $p > k$ and replace K^k by the kernel of the differential $K^k \to K^{k+1}$, while in $\tau_{\geq k}K$ all K^p with $p < k$ are removed while K^k is replaced by the co-image of $K^{k-1} \to K^k$. The crucial property of the complex $\tau_{\leq k}K$ is that it has the same cohomology as K in degrees $\leq k$ but no higher cohomology (and similarly for $\tau_{\geq k}K$). The operation $\tau_{\leq 0}$ followed by $\tau_{\geq 0}$ thus produces out of K a complex with cohomology concentrated in degree 0 only. In general, one has

$$H^k(K) = \tau_{\leq 0}\tau_{\geq 0}(K[k]),$$

viewed as a complex in degree zero. This procedure enables us to recover the abelian category $\mathfrak{A}$ from its derived category $D(\mathfrak{A})$: upper truncation followed by lower truncation at degree 0 identifies $\mathfrak{A}$ with the subcategory of complexes concentrated in degree zero.

These constructions can be understood abstractly within the framework of triangulated categories; this leads to the notions of t-structure and heart. Before we discuss these, let us explain where distinguished triangles originate from. First recall:

DEFINITION 8.2.4. Let $A^\bullet \xrightarrow{f} B^\bullet$ be a morphism of complexes in an additive category $\mathfrak{A}$. The *cone* of f, denoted $\mathrm{Cone}(f)$, is the complex $A^\bullet[1] \oplus B^\bullet$ with differential $d(a, b) = (-da, db + f(a))$.

One can show that the short exact sequence $0 \to A \xrightarrow{f} B \to C \to 0$ can be replaced by $0 \to A \xrightarrow{f} B' \to \mathrm{Cone}(f) \to 0$ with B' homotopy equivalent to B. So, up to homotopy we can replace C by $\mathrm{Cone}(f)$; furthermore, in the homotopy category we can permute the three objects $A, B, \mathrm{Cone}\, f$ cyclically: we get a so-called *distinguished triangle* $A \xrightarrow{f} B \to \mathrm{Cone}(f) \to A[1]$ in $D(\mathfrak{A})$. Such a triangle encodes all that is needed to write down the long exact sequence in cohomology. This will be generalized in the concept of a triangulated category.

DEFINITION. A *triangulated category* is an additive category D such that
   (1) there is a a shift operator $A \mapsto A[1]$;
   (2) there are distinguished triangles $A \to B \to C \to A[1]$ satisfying certain axioms first formulated by Verdier; see [**Verd**].

---

[2]Recall that for any complex K in an abelian category the complex $K[d]$ denotes the complex shifted d places to the left, i.e., $K[d]^i = K^{d+i}$.

Next, a *t-structure* on a triangulated category D consists of

(1) two full subcategories $D^{\leq 0}$ and $D^{\geq 0}$ which correspond to complexes having only cohomology in degree ≤ 0 and ≥ 0 respectively. The translation functor then makes it possible to define $D^{\leq k}$ as consisting of objects $K[-k]$, $K \in D^{\leq 0}$ and similarly $D^{\geq k} = \{K[-k] \mid K \in D^{\geq 0}\}$. Some obvious inclusion relations must hold: $D^{\leq -1} \subset D^{\leq 0}$ and $D^{\geq 1} \subset D^{\geq 0}$;

(2) truncation functors $\tau_{\leq 0} : D \mapsto D_{\leq 0}$ and $\tau_{\geq 1} : D \mapsto D_{\geq 1}$ such that $\tau_{\leq 0}K \to K \to \tau_{\geq 1}K \to \tau_{\leq 0}K[1]$ is a distinguished triangle.

The *heart* of a t-structure is defined as $c(D) = D^{\leq 0} \cap D^{\geq 0}$. The heart can be proven to be an abelian category.

To introduce cohomology, we need to define the other truncation functors of a t-structure:

$$\tau_{\geq k}K = \tau_{\geq 1}(K[k-1])[-k+1], \quad \tau_{\leq k}(K) = \tau_{\leq 0}(K[k])[-k].$$

The cohomology associated to the t-structure is defined by

$$^tH^0(K) := \tau_{\leq 0}\tau_{\geq 0}K, \qquad {}^tH^k(K) := \tau_{\leq 0}\tau_{\geq 0}(K[k]).$$

By construction $^tH^k(K)$ belong to the heart of D.

One applies these considerations to $D = D^b_c(\mathbb{Q}_S)$. Define $D^{\leq 0}$ as the category consisting of the complexes in this category satisfying the support condition while $D^{\geq 0}$ are those which satisfy the co-support condition. Indeed, one has:

THEOREM ([**Beil-Ber-Del**, Corollary 2.1.4]). *The support and cosupport conditions define a t-structure, the so-called* perverse t-structure, *on* $D^b_c(\mathbb{Q}_S)$. *Hence* $\mathrm{Perv}(S)$, *being the heart of this perverse t-structure, is an abelian category.*

The truncation functors of the perverse t-structure are denoted $^p\tau_{\leq 0}$ and $^p\tau_{\geq 0}$, and its cohomology functors are called the *perverse cohomology functors*. Explicitly

$$\begin{aligned} ^pH^k : D^b_c(\mathbb{Q}_S) &\to \mathrm{Perv}(S) \\ F^\bullet &\mapsto {}^p\tau_{\leq 0}{}^p\tau_{\geq 0}(F^\bullet[k]). \end{aligned}$$

Given $(X, f) \in \mathsf{Var}(S)$, and $F = F^\bullet$ perverse on X, there is a special notation for the perverse direct image sheaf on Y:

$$^pR^i f_*(F) := {}^pH^i(Rf_*F)),$$

which is motivated by the fact, recalled before (see equation (47)), that the usual direct image $R^i f_* F$ equals the cohomology sheaf $\mathcal{H}^i(Rf_*F)$.

EXAMPLE 8.2.5. If $f : X \to S$ is a smooth morphism and S is smooth, the sheaves $R^i f_* \mathbb{Q}$ are local systems and $(R^i f_* \mathbb{Q})[d_S]$ is a perverse sheaf for all i (see Example 8.2.3). In this case we have isomorphisms of perverse sheaves (cf. [**Ca-Mi09**, Remark 1.5.1])

$$^pR^i f_*(\mathbb{Q}[d_X]) \simeq (R^{i+d_X-d_S} f_* \mathbb{Q})[d_S].$$

THEOREM 8.2.6 (Decomposition theorem, [**Beil-Ber-Del**, Theorem 6.2.5]). *Suppose that X is smooth and that $f : X \to S$ is a projective morphism. Then*

$$Rf_*(\mathbb{Q}_X[d_X]) \simeq \oplus_i {}^pR^i f_*(\mathbb{Q}_X[d_X])[-i] \tag{52}$$

$$^pR^i f_*(\mathbb{Q}_X[d_X]) \simeq \oplus_Z \mathrm{IC}_Z(\mathbb{V}^i_{Z_0}), \tag{53}$$

where the sum runs over the irreducible subvarieties $Z \subset S$, $Z_0 \subset Z$ is a dense smooth open subset, $\mathbb{V}^i_{Z_0}$ is a local sytem on Z_0 and $\{Z \mid \mathbb{V}^i_{Z_0} \neq 0\}$ is a finite set.

Remark 8.2.7. If f is a stratified map (cf.[**Pe-St**, Appendix C.1.2]), the subvarieties Z can be taken to be closures of strata; cf. [**Cor-Ha00, Gor-Ha-Mu**]. The survey article [**Ca-Mi09**] contains a very nice discussion of the decomposition theorem.

8.3. Relative Chow-Künneth Decomposition

8.3.1. Realization Functor. For possibly singular varieties not only are Fulton's co-variant Chow groups better behaved, also instead of cohomology, it is better to use co-variantly behaving Borel-Moore homology. For instance if S is smooth but not complete, the Borel-Moore group $H_i^{\mathrm{BM}}(S)$ with $\mathbb{Q}$–coefficients is the same as ordinary cohomology $H^{2d_S-i}(S)$.

We next consider the cycle class map for Borel-Moore homology:

$$\mathsf{Corr}_S^p(X,Y) = \mathsf{CH}_{d_Y-p}(X \times_S Y)_{\mathbb{Q}} \to H_{2d_Y-2p}^{\mathrm{BM}}(X \times_S Y, \mathbb{Q}).$$

We want to rewrite the right hand side as a group of homomorphisms so that we can map a correspondence to such a homomorphism:

LEMMA 8.3.1 ([**Cor-Ha00**, Lemma 2.21 (2)]). *Let* (X,f), $(Y,g) \in \mathsf{Var}(S)$. *We have*

$$H_{2d_Y+i-j}^{\mathrm{BM}}(X \times_S Y, \mathbb{Q}) \cong \mathrm{Hom}_{D_c^b(\mathbb{Q}_S)}(Rf_*\mathbb{Q}_X[i], Rg_*\mathbb{Q}_Y[j]).$$

So, indeed, we deduce:

COROLLARY 8.3.2. *There exists a realization map*

$$\psi : \mathsf{Corr}_S^p(X,Y) \to \mathrm{Hom}(Rf_*\mathbb{Q}_X, Rg_*\mathbb{Q}_Y[2p]).$$

In particular, the cycle class map $\gamma_{X \times_S X} : \mathsf{CH}_{d_X}(X \times_S X) \to H_{2d_X}^{\mathrm{BM}}(X \times_S X)$ *corresponds to a map*

$$\mathrm{End}_{\mathsf{Mot}_{\mathrm{rat}}(S)}(\mathsf{ch}_S(X)) = \mathsf{Corr}_S^0(X,X) \to \mathrm{End}_{D_c^b(\mathbb{Q}_S)}(Rf_*\mathbb{Q}_X).$$

Given $\Gamma \in \mathsf{Corr}_S^p(X,Y)$, we write

$$\Gamma_* : Rf_*\mathbb{Q}_X \to Rg_*\mathbb{Q}_Y[2p]$$

for the induced map in $D_c^b(\mathbb{Q}_S)$. As $D_c^b(\mathbb{Q}_S)$ is pseudo–abelian ([**Cor-Ha00**, Lemma 2.24], [**Bal-Sc, Le-C**]), projectors have images. Hence we can extend this construction to relative motives. In this way we obtain a *realization functor*

$$\mathsf{real} : \mathsf{Mot}_{\mathrm{rat}}(S) \to D_c^b(\mathbb{Q}_S)$$

which sends $((X,f), p(X/S), n)$ to $p(X/S)_* Rf_*\mathbb{Q}_X[2n]$.

8.3.2. Motivic Decomposition Conjecture.

DEFINITION 8.3.3. A projective morphism $f : X \to S$ admits a *relative Chow-Künneth decomposition in the weak sense* if there exist mutually orthogonal relative projectors $p_i(X/S)$ that induce the decomposition (52) in $D_c^b(\mathbb{Q}_S)$, i.e.,

  (i) $\sum_i p_i(X/S) = \Delta_X$;
  (ii) $p_i(X/S) \circ p_j(X/S) = \delta_{ij} p_i(X/S)$;
  (iii) $p_i(X/S)_*|_{^p R^j f_*(\mathbb{Q}_X[d_X])} = \begin{cases} 0 & i \neq j \\ \mathrm{id} & i = j. \end{cases}$

We say that $f : X \to S$ admits a *relative Chow-Künneth decomposition in the strong sense* if there exists mutually orthogonal relative projectors $p_{i,Z}(X/S)$ that induce the decomposition (53), i.e.,

(i) $\sum_i p_{i,Z}(X/S) = \Delta_X$;

(ii) $p_{i,Z}(X/S) \circ p_{j,W}(X/S) = \left\{ \begin{array}{ll} 0 & \text{if } (i,Z) \neq (j,W) \\ p_{i,Z}(X/S) & \text{if } (i,Z) = (j,W); \end{array} \right.$

(iii) $p_{i,Z}(X/S)_* |_{\mathrm{IC}_W(\mathbb{V}^j_{W^0})} = \left\{ \begin{array}{ll} 0 & (i,Z) \neq (j,W) \\ \mathrm{id} & (i,Z) = (j,W). \end{array} \right.$

Remark. If f is a smooth morphism, the notions of Chow-Künneth decomposition in the weak and strong sense coincide.

CONJECTURE 8.3.4 (Motivic Decomposition Conjecture [**Cor-Ha00**], [**Gor-Ha-Mu**]). For every $(X,f) \in \mathsf{Var}(S)$, the morphism $f : X \to S$ admits a relative Chow-Künneth decomposition in the strong sense.

PROPOSITION 8.3.5. *If the cycle class map*

$$\gamma : \mathsf{CH}_{d_X}(X \times_S X)_\mathbb{Q} \to H^{\mathrm{BM}}_{2d_X}(X \times_S X; \mathbb{Q})$$

is an isomorphism, then $X \to S$ admits a relative Chow-Künneth decomposition (in the strong sense).

Proof: By Corollary 8.3.2, γ corresponds to an isomorphism

$$\mathrm{real} : \mathsf{Corr}^0_S(X,X) \xrightarrow{\simeq} \mathrm{End}(Rf_*\mathbb{Q}_X).$$

Choose projectors $p_i^{\mathrm{top}}(X/S) \in \mathrm{End}(Rf_*\mathbb{Q}_X)$ that induce the decomposition (53), and lift them to projectors $p_i(X/S) \in \mathsf{Corr}^0_S(X,X)$. The projectors $p_i(X/S)$ give a relative Chow-Künneth decomposition (in the strong sense) of $X \to S$. $\qquad\square$

EXAMPLE 8.3.6 (Semi-small Maps). Recall that a proper, surjective map $f : X \to S$ is called *semi-small* if

$$\dim\{s \in S \mid \dim f^{-1}(s) = k\} \leq \dim X - 2k$$

for all k. (This implies in particular that the map f is generically finite.) The map being semi-small implies that every irreducible component of $X \times_S X$ has dimension $\leq d_X$, hence the cycle class map

$$\gamma : \mathsf{CH}_{d_X}(X \times_S X)_\mathbb{Q} \to H^{\mathrm{BM}}_{2d_X}(X \times_S X)$$

is an isomorphism. Proposition 8.3.5 implies that $f : X \to S$ admits a relative Chow-Künneth decomposition in the strong sense.

8.4. Further Examples

Throughout this section we consider the following special situation, which was studied in [**Den-Mu**]. Let S be *smooth* quasi-projective variety, and consider the category $\mathcal{V}(S)$ of varieties X equipped with a *smooth*, projective morphism $f : X \to S$. In this case, the formula (49) simplifies: given two relative correspondences $\Gamma_1 \in \mathsf{Corr}^p_S(X,Y)$ and $\Gamma_2 \in \mathsf{Corr}^q_S(Y,Z)$ we have

$$\Gamma_2 \circ_S \Gamma_1 := (p_{XZ})_*(p_{XY}^*(\Gamma_1) \cdot p_{YZ}^*(\Gamma_2)) \in \mathsf{Corr}^{p+q}_S(X,Z)$$

where $p_{XZ} : X \times_S Y \times_S Z \to X \times_S Z$ denotes the projection map, and similarly for p_{XY}, p_{YZ}. In this setting we have the following relative version of the Manin identity principle. Given $T \in \mathcal{V}(S)$, put $X_S(T) := \mathsf{Corr}_S(T,X) = \mathsf{CH}(X \times_S T; \mathbb{Q})$ and for $f \in \mathsf{Corr}_S(X,Y)$, let

$$f_T : X_S(T) \to Y_S(T), \quad \alpha \mapsto f \circ \alpha.$$

Then we have the:

PROPOSITION 8.4.1 (Relative Manin identity principle). *Let $f, g \in \mathsf{Corr}_S(X, Y)$. Then the following are equivalent:*

    (1) $f = g$;
    (2) $f_T = g_T$ *for all* $T \in \mathcal{V}(S)$;
    (3) $f_X = g_X$.

In particular,

> f *is an isomorphism* $\iff$ f_T *is an isomorphism for all* T $\iff$ f_X *is an isomorphism.*

As before, there is a relative version of Lieberman's lemma which implies the following result.

COROLLARY 8.4.2. *Let $f, g \in \mathsf{Corr}_S(X, Y)$. We have $f = g$ as relative correspondences if and only if*

$$(\mathrm{id}_T \times_S f)_* = (\mathrm{id}_T \times_S g)_* : \mathsf{CH}(X \times_S T) \to \mathsf{CH}(Y \times_S T)$$

for all $T \in \mathcal{V}(S)$.

8.4.1. Abelian Schemes over a Base. Let $f : A \to S$ be an abelian scheme over S of relative dimension g. For every integer n, we have a map $\mathrm{id}_A \times n : A \times_S A \to A \times_S A$. Using the theory of the Fourier transform, Deninger and Murre showed [**Den-Mu**] that there exist mutually orthogonal projectors $p_i(A/S)$ such that

$$\Delta_A = \sum_{i=0}^{2g} p_i(A/S) \tag{54}$$

and such that $(\mathrm{id}_A \times n)^* p_i(A/S) = n^i p_i(A/S)$ for all $n \in \mathbb{Z}$. The projectors $p_i(A/S)$ give a relative Chow-Künneth decomposition of A over S.

8.4.2. Morphisms that Admit a Relative Cell Decomposition. A morphism $f : X \to S$ admits a *relative cell decomposition* if there exists a filtration

$$X = X_0 \supset X_1 \supset \ldots \supset X_n \supset \varnothing$$

of closed subschemes $X_\alpha \subset X$ such that for all α

    (1) $f_\alpha = f|_{X_\alpha} : X_\alpha \to S$ is flat;
    (2) the open subset $U_\alpha = X_\alpha - X_{\alpha+1}$ is an affine bundle over S under

$$\pi_\alpha := f|_{U_\alpha} \to S.$$

PROPOSITION 8.4.3. *Suppose $f : X \to S$ admits a relative cell decomposition. Then the relative Chow motive is a direct sum of relative Lefschetz-motives and f admits a relative Chow-Künneth decomposition.*

Proof: Let $j_\alpha : U_\alpha \hookrightarrow X_\alpha$ and $i_\alpha : X_{\alpha+1} \hookrightarrow X_\alpha$ be the inclusions. Suppose that the codimension of $X_{\alpha+1}$ in X_α equals d_α and consider the localization sequence

$$\mathsf{CH}^{k-d_\alpha}(X_{\alpha+1})_\mathbb{Q} \xrightarrow{(i_{\alpha+1})_*} \mathsf{CH}^k(X_\alpha)_\mathbb{Q} \xrightarrow{j_\alpha^*} \mathsf{CH}^k(U_\alpha)_\mathbb{Q} \to 0.$$

One has $(\pi_\alpha)^* : \mathsf{CH}^k(S)_\mathbb{Q} \xrightarrow{\sim} \mathsf{CH}^k(U_\alpha)_\mathbb{Q}$ since π_α is an affine bundle. Consider the closure Γ_α in $X_\alpha \times S$ of the graph of π_α. Its transpose is a degree zero correspondence from S to X_α. Then

$$\mathsf{CH}^k(U_\alpha)_\mathbb{Q} \xrightarrow{(\pi_\alpha^*)^{-1}} \mathsf{CH}^k(S)_\mathbb{Q} \xrightarrow{{}^{\mathrm{T}}\Gamma_\alpha} \mathsf{CH}^k(X_\alpha)_\mathbb{Q}$$

is easily seen to be a section of the morphism $j_\alpha^* : \mathsf{CH}^k(X_\alpha)_\mathbb{Q} \to \mathsf{CH}^k(U_\alpha)_\mathbb{Q}$. Let $k_\alpha : X_\alpha \hookrightarrow X$ be the inclusion and let $c_\alpha = d_\alpha + d_{\alpha-1} + \cdots + d_1$ be the codimension of X_α in X. Using induction it thus follows that

$$\oplus_\alpha \gamma_\alpha^* : \bigoplus_\alpha \mathsf{CH}^{k-c_\alpha}(S)_\mathbb{Q} \xrightarrow{\sim} \mathsf{CH}^k(X)_\mathbb{Q}, \quad \gamma_\alpha := k_\alpha \circ {}^\mathsf{T}\Gamma_\alpha \in \mathsf{Corr}^{c_\alpha}(S, X),$$

an isomorphism, which is moreover functorial with respect to cartesian squares

$$\begin{array}{ccc} X \times_S T & \to & X \\ \downarrow {\scriptstyle f_T} & & \downarrow {\scriptstyle f} \\ T & \to & S. \end{array}$$

Hence, using the relative version of Manin's identity principle (Corollary 8.4.2) and (51), we obtain

$$\mathsf{ch}_S(X) \cong \bigoplus_\alpha \mathbf{L}_S^{\otimes c_\alpha}.$$

This decomposition is given by a set of mutually orthogonal projectors $p_\alpha(X/S) \in \mathsf{Corr}_S^0(X, X)$. As the maps $p_\alpha(X/S)_*$ induce the decomposition

$$Rf_* \mathbb{Q}_X \simeq \oplus_\alpha \mathbb{Q}_S[-2c_\alpha]$$

in $D_c^b(\mathbb{Q}_S)$, the result follows. $\qquad\square$

8.5. From Relative to Absolute

Given a smooth, projective k-variety X that is fibered over a base S, one could try to obtain a Chow-Künneth decomposition of X in two steps:

1) Find relative projectors $p_i(X/S)$ that give a relative Chow-Künneth decomposition of X over S (in the strong sense);
2) Take the images $p_i(X)$ of the projectors $p_i(X/S)$ under the map $\mathsf{Mot}_{\mathrm{rat}}(S) \to \mathsf{Mot}_{\mathrm{rat}}(k)$; these are projectors by Lemma 8.1.6 (2). Try to construct an absolute Chow-Künneth decomposition of X using the projectors $p_i(X)$, by decomposing these further if necessary.

This approach was initiated in [**Gor-Ha-Mu**]. In the previous section, we have seen some examples where Step 1 can be carried out. The problem in Step 2 is that the motives $M_i = (X, p_i)$ may have cohomology in more than one degree; in this case, one should find a decomposition $M_i = \oplus_j M_{ij}$ of the motive M_i such that M_{ij} has cohomology in only one degree.

We shall illustrate this method in the following example.

EXAMPLE 8.5.1 (Conic bundles over a surface [**Nag-Sa**]). Let k be an algebraically closed field of characteristic zero, and let $f : X \to S$ be a conic bundle over a surface, i.e., X is a smooth projective threefold defined over k, S is a smooth projective surface over k and the fiber $f^{-1}(s)$ is a conic for all $s \in S$. Let $C \subset S$ be the discriminant curve parametrising the singular fibers, and let $\rho : \widetilde{C} \to C$ be the double covering parametrising the irreducible components of the singular fibers.

For simplicity, we shall make a few additional assumptions[3]:

  (i) all the fibers of f are reduced (i.e. double lines do not occur)
  (ii) ρ is a nontrivial double covering
  (iii) C is connected.

---

[3]As shown in [**Nag-Sa**] these assumption are superfluous.

If these assumptions are satisfied, C and $\widetilde{C}$ are smooth and irreducible, and ρ is an étale double covering; cf. [**Beau77**, Propositions 1.2 and 1.5].

Step 1: Relative Chow-Künneth Decomposition.   Let us start by describing the result of the two decompositions (52), (53) in our example:

PROPOSITION 8.5.2. *We have*[4]

$$Rf_*(\mathbb{Q}_X[2]) \simeq \mathbb{Q}_S[2] \oplus \mathbb{Q}_S \oplus i_*\mathbb{L}.,$$

where $\mathbb{L}$ is a rank one local system on C, and $i : C \to S$ the inclusion.

Proof: By [**Beau77**, Prop. 1.2] the conic bundle X embeds into a projective bundle $Y = \mathbb{P}(E) \xrightarrow{\varphi} S$ with E a rank three vector bundle over S. Using the relative version of the Lefschetz hyperplane theorem for perverse sheaves (see e.g. [**Ca-Mi09**, Thm. 2.6.4]), the smoothness of φ and Example 8.2.5 we obtain

$${}^p R^{-1} f_*(\mathbb{Q}_X[3]) \simeq {}^p R^{-2} \varphi_*(\mathbb{Q}_Y[4]) \simeq (R^0 \varphi_* \mathbb{Q}_X)[2] \simeq \mathbb{Q}_S[2].$$

The relative hard Lefschetz theorem [**Beil-Ber-Del**] implies that ${}^p R^1 f_*(\mathbb{Q}_X[3]) \simeq {}^p R^{-1} f_*(\mathbb{Q}_X[3]) \simeq \mathbb{Q}_S[2]$. Hence the decomposition theorem of [**Beil-Ber-Del**] gives an isomorphism

$$Rf_*(\mathbb{Q}_X[3]) \simeq \mathbb{Q}_S[3] \oplus ({}^p R^0 f_*(\mathbb{Q}_X[3]) \oplus \mathbb{Q}_S[1].$$

Comparing the cohomology sheaves on both sides, we obtain ${}^p R^0 f_*(\mathbb{Q}_X[3]) \simeq i_*\mathbb{L}[1]$ in $\mathrm{Perv}(S)$, with $\mathbb{L} = \mathrm{Coker}(R^2 \varphi_*(\mathbb{Q}_X) \to R^2 f_*(\mathbb{Q}_Y))$, a rank one local system supported on C. Shifting one unity to the right gives the desired decomposition. $\square$

The conic bundle $X \to S$ admits a multi-section, i.e., there exists $Z \subset X$ such that $Z \to S$ is a double covering. Put $\xi = \frac{1}{2}[Z] \in \mathsf{CH}^1(X)$, and define

$$p_0'(X/S) = \xi \times_S X, \quad p_2'(X/S) = X \times_S \xi.$$

By [**Nag-Sa**, Lemma 2.1] we have

$$p_0'(X/S) = [X] \circ \xi$$

with $[X] \in \mathsf{CH}^0(X) \cong \mathsf{Corr}_S^0(S, X)$ and $\xi \in \mathsf{CH}^1(X)_{\mathbb{Q}} \cong \mathsf{Corr}_S^0(X, S)$. By transposition $p_2'(X/S) = {}^\mathsf{T}\xi \circ {}^\mathsf{T}[X]$. As $f_*(\xi) = [S]$, we have $\xi \circ [X] = \mathrm{id} = {}^\mathsf{T}[X] \circ {}^\mathsf{T}\xi$, hence $p_0'(X/S)$ and $p_2'(X/S)$ are projectors.

We have $p_2'(X/S) \circ p_0'(X/S) = 0$, but $p_0'(X/S) \circ p_2'(X/S)$ need not be zero, as the cycle ξ may have a nontrivial self-intersection. We correct this by putting

$$
\begin{aligned}
p_0(X/S) &= p_0'(X/S) - \tfrac{1}{2}p_0'(X/S) \circ p_2'(X/S) & (55) \\
p_2(X/S) &= p_2'(X/S) - \tfrac{1}{2}p_0'(X/S) \circ p_2'(X/S) & (56)
\end{aligned}
$$

to obtain orthogonal projectors. Put $p_1(X/S) = \mathrm{id} - p_0(X/S) - p_2(X/S)$.

The projector $p_0'(X/S)$ acts as $\delta_{0j} \cdot \mathrm{id}$ on ${}^p R^j f_*\mathbb{Q}$ since its restriction to every smooth fiber $X_s = f^{-1}(s)$ is the projector $p_0(X_s)$ constructed in Example 2.3 (ii). Similarly, $p_2'(X/S)$ acts as $\delta_{2j} \cdot \mathrm{id}$ on ${}^p R^j f_*\mathbb{Q}$. By (55) and (56) the projectors $p_i(X/S)$ act as $\delta_{ij} \cdot \mathrm{id}$ on ${}^p R^j f_*\mathbb{Q}$ for $i = 0, 2$. Since the remaining projector $p_1(X/S)$ is supported on C, it acts as zero on ${}^p R^j f_*\mathbb{Q}$ for $j = 0, 2$, and hence as the identity on ${}^p R^1 f_*\mathbb{Q}$. So the projectors $p_i(X/S)$ give a relative Chow-Künneth decomposition (in the strong sense).

---

[4]According to the convention of Theorem 8.2.6 we should consider $Rf_*(\mathbb{Q}_X[3])$. Since we apply the result to the Chow-Künneth conjecture, it is more practical to use a different shift which results in sheaves placed in degrees 0, 1 and 2 in stead of -1, 0 and 1.

Step 2: Absolute Chow-Künneth decomposition. Let $p_i(X)$ be the image of $p_i(X/S)$ under the map $\mathsf{Mot}_{\mathrm{rat}}(S) \to \mathsf{Mot}_{\mathrm{rat}}(k)$ $(i = 0, 1, 2)$. Put $M_i = (X, p_i)$. The correspondence $\xi \in \mathsf{Corr}_S^0(X, S)$ induces an isomorphism $M_0 \cong \mathsf{ch}(S)$ with inverse $[X] \in \mathsf{Corr}_S^0(X, S)$ (This follows from the relation $\xi \circ [X] = \mathrm{id}$.) The correspondences ${}^\mathsf{T}[X]$ and ${}^\mathsf{T}\xi$ induce an isomorphism $M_2 \cong \mathsf{ch}(S)(-1)$.

The decomposition

$$\mathsf{ch}(S) = \mathsf{ch}^0(S) + \mathsf{ch}^1(S) + \mathsf{ch}^2(S) + \mathsf{ch}^3(S) + \mathsf{ch}^4(S)$$

gives decompositions

$$
\begin{aligned}
M_0 &= M_{00} + M_{01} + M_{02} + M_{03} + M_{04} \\
M_2 &= M_{20} + M_{21} + M_{22} + M_{23} + M_{24}
\end{aligned}
$$

such that the motives M_{ij} have cohomology in only one degree.

It remains to consider the cohomology of the motive M_1. As $p_{1,*}(Rf_*\mathbb{Q}_X) \cong i_*\mathbb{L}(-1)[-2]$ we have

$$H^i(M_1) \cong H^{i-2}(C, \mathbb{L}).$$

By assumption the local sytem $\mathbb{L}$ is nontrivial, hence $H^0(C, \mathbb{L}) = 0$. By Poincaré-Verdier duality $H^2(C, \mathbb{L}) \cong H^0(C, \mathbb{L}^\vee) = 0$. Hence M_1 only has cohomology in degree three, and we obtain an absolute Chow-Künneth decomposition

$$\mathsf{ch}(X) = \mathsf{ch}^0(X) + \mathsf{ch}^1(X) + \mathsf{ch}^2(X) + \mathsf{ch}^3(X) + \mathsf{ch}^4(X) + \mathsf{ch}^5(X) + \mathsf{ch}^6(X)$$

with

$$
\begin{aligned}
\mathsf{ch}^0(X) &\cong \mathsf{ch}^0(S) \\
\mathsf{ch}^1(X) &\cong \mathsf{ch}^1(S) \\
\mathsf{ch}^2(X) &\cong \mathsf{ch}^0(S)(-1) + \mathsf{ch}^2(S) \\
\mathsf{ch}^3(X) &\cong \mathsf{ch}^1(S)(-1) + \mathsf{ch}^3(S) + M_1 \\
\mathsf{ch}^4(X) &\cong \mathsf{ch}^2(S)(-1) + \mathsf{ch}^4(S) \\
\mathsf{ch}^5(X) &\cong \mathsf{ch}^3(S)(-1) \\
\mathsf{ch}^6(X) &\cong \mathsf{ch}^4(S)(-1).
\end{aligned}
$$

Remark. For any étale double cover $\sigma : \widetilde{C} \to C$ of smooth projective curves, the Prym-projector is the projector $p_{\mathrm{prym}} := \frac{1}{2}(\mathrm{id} - \sigma^*) \in \mathsf{Corr}_C^0(\widetilde{C}, \widetilde{C})$ and the corresponding Prym-motive is

$$\mathrm{Prym}(\widetilde{C}/C) := (\widetilde{C}, p_{\mathrm{prym}}).$$

It can be shown that the motive M_1 is isomorphic to $\mathrm{Prym}(\widetilde{C}/C)(-1)$; this is the main geometric input of [**Nag-Sa**].

Appendix D: Surfaces Fibered over a Curve

Let S be a smooth projective surface that admits a surjective, generically smooth morphism $f : S \to C$ to a smooth, projective curve C. Let $i : \Sigma \hookrightarrow C$ be the discriminant locus, which is a finite set. Put $S_\Sigma = f^{-1}(\Sigma) = \sqcup_{t\in\Sigma} S_t$, let $\widetilde{S_\Sigma} \to S_\Sigma$ the normalization and let $\iota : \widetilde{S_\Sigma} \to S$ the obvious map. We put $U = C - \Sigma$ and $S_U = f^{-1}U$ and we let $j : U \hookrightarrow C$ stand for the natural inclusion. These maps figure in the diagram

$$
\begin{array}{ccccc}
S_U & \hookrightarrow & S & \xleftarrow{\iota} & \widetilde{S_\Sigma} \\
{\scriptstyle f|U}\downarrow & & {\scriptstyle f}\downarrow & & \downarrow \\
U & \hookrightarrow & C & \xleftarrow{i} & \Sigma.
\end{array}
$$

THEOREM 8.5.3. *In the above situation, assume that*

(i) *the singular fibers $S_t = f^{-1}(t)$, $t \in \Sigma$, are divisors with strict normal crossings. Hence the components are smooth and have multiplicity 1.*
(ii) *the morphism $f : S \to C$ admits a section $\varepsilon : C \to S$. We let $B = \varepsilon(C) \subset S$.*

Then $f : S \to C$ admits a strong relative Chow-Künneth decomposition.

The remainder of this Appendix is devoted to a sketch of the proof of this Theorem. We use the following convention: $S_{t,o}$ is the component of the fibre S_t meeting the section and we write

$$
S_t = S_{t,o} + \sum_{\alpha \neq o} S_{t,\alpha} \quad t \in \Sigma.
$$

Start with the decomposition from Theorem 8.2.6. As explained in footnote 4, it is better to perform a shift to the right; one then gets (see for example [**Ca-Mi07**, § 3.3.2]):

$$
Rf_*(\mathbb{Q}_S[1]) \simeq \underbrace{\mathbb{Q}_C[1]}_{{}^p R^0 f_*(\mathbb{Q}_S[1])} \oplus \underbrace{\mathbb{Q}_C[-1]}_{{}^p R^2 f_*(\mathbb{Q}_S[1])[-2]} \oplus \underbrace{IC_C(R^1 f_{U,*}\mathbb{Q})[-1] \oplus i_*\mathbb{V}[-1]}_{{}^p R^1 f_*(\mathbb{Q}_S[1])[-1]}, \quad (57)
$$

where

$$
\mathbb{V} := \mathrm{Ker}\left[R^2 f_*\mathbb{Q} \to j_*(R^2 f_*\mathbb{Q}|U)\right]
$$

which is a skyscraper sheaf on Σ whose fiber over $t \in \Sigma$ is a $\mathbb{Q}$–vector space of dimension one less the number of components of S_t. Explicitly, we have

$$
\mathbb{V}_t = \left[\bigoplus_\alpha \mathbb{Q} \cdot \gamma_{\widetilde{S}_{t,\alpha}}(\widetilde{S}_{t,\alpha})\right] / \mathbb{Q} \cdot \gamma_{\widetilde{S}_{t,\alpha}}(\iota^* S_t). \quad (58)
$$

As we did before (see formulas (55) and (56)), we modify the projectors

$$
p'_0(S/C) = B \times_C S, \quad p'_2(S/C) = S \times_C B
$$

to obtain orthogonal projectors

$$p_0(S/C) = p_0'(S/C) - \tfrac{1}{2}p_0'(S/C)p_2'(S/C) \tag{59}$$

$$p_2(S/C) = p_2'(S/C) - \tfrac{1}{2}p_0'(S/C)p_2'(S/C) \tag{60}$$

that induce the projections from $Rf_*\mathbb{Q}_S$ to the first two factors in the above decomposition (57).

We now look for a relative projector

$$p_\infty(S/C) \in \mathsf{Corr}_C^0(S,S) = \mathsf{CH}_2(S \times_C S)_\mathbb{Q}$$

that induces the projection to $i_*\mathbb{V}[-1]$. As $\widetilde{S}_\Sigma \times_\Sigma \widetilde{S}_\Sigma$ is two-dimensional, the cycle class map

$$
\begin{array}{ccc}
\mathsf{CH}_2(\widetilde{S}_\Sigma \times_\Sigma \widetilde{S}_\Sigma)_\mathbb{Q} & \xrightarrow{\;\gamma_{\widetilde{S}_\Sigma \times_\Sigma \widetilde{S}_\Sigma}\;} & H_4(\widetilde{S}_\Sigma \times_\Sigma \widetilde{S}_\Sigma) \\
\| & & \| \\
\oplus_{t\in\Sigma}\mathsf{CH}_2(\widetilde{S}_t \times \widetilde{S}_t)_\mathbb{Q} & \xrightarrow{\;\sim\;} & \oplus_{t\in\Sigma}H_4(\widetilde{S}_t \times \widetilde{S}_t)
\end{array}
\tag{61}
$$

is an isomorphism. Put

$$p_\infty(S/C) := \sum_{t\in\Sigma}\sum_{\alpha,\beta\neq o} m_{\alpha,\beta}S_{t,\alpha} \times S_{t,\beta},$$

where $(m_{\alpha,\beta})^{-1}$ is the intersection matrix of the components $S_{t,\alpha}$, $\alpha \neq 0$. We view this as an element in $\mathsf{CH}_2(S \times_C S)_\mathbb{Q}$. As can be easily seen directly from the definitions, it is a projector (same argument as Appendix C) and through the cycle class map it acts as the projection onto the last factor $i_*\mathbb{V}[-1]$ in the above decomposition (57), where the reader should bear in mind the cycle class isomorphism (61) (compare [**Cor-Ha07**, Proposition 1.5], [**Gor-Ha-Mu**, Prop. I-3]).

It remains to show that $p_\infty(S/C)$ is orthogonal to $p_0(S/C)$ and $p_2(S/C)$. By (55) and (56) it suffices to show that $p_\infty(S/C)$ is orthogonal to p_0' and p_2'. From the definition of $p_\infty(S/C)$ we clearly have $p_\infty = {}^\mathsf{T}p_\infty$. So, as p_2' is the transpose of p_0', it suffices to show that

$$p_\infty \circ p_0' = p_0' \circ p_\infty = 0.$$

Now, by definition, to calculate $p_0' \circ p_\infty$, we need to form the intersection product $(p_\infty \times_C S) \cdot (S \times_C B \times_C S)$ and then apply $(p_{13})_*$ to the result, where p_{13} means projection onto the product of the first and third factor. Since B does not meet any of the $S_{t,\beta}$, $\beta \neq 0$, the above intersection product is trivially zero and a fortiori we have $p_0' \circ p_\infty = 0$. To prove that $p_\infty \circ p_0' = 0$ we have to apply $(p_{13})_*$ to the intersection product $Z := (B \times_C S \times_C S).(S \times_C p_\infty)$. This also vanishes since a dimension count shows that $\dim Z = 2$ (Z fibers over Σ with 2-dimensional fibers) and $\dim p_{13}(Z) = 1$, hence $(p_{13})_*(Z) = 0$ by definition. Finally we put

$$p_1(S/C) = \mathrm{id} - p_0(S/C) - p_2(S/C) - p_\infty(S/C).$$

By construction $p_1(S/C)_*$ acts as the identity on $\mathrm{IC}_C(R^1 f_{U,*}\mathbb{Q})$ and is zero on the other factors. Hence the projectors $p_i(S/C)$ ($i = 0,1,2,\infty$) give a relative Chow-Künneth decomposition of $S \to C$ in the strong sense.

Remark 8.5.4. To pass from a relative to an absolute Chow-Künneth decomposition in this case, one should split the motive $M_1 = (S, p_1)$ into a 'constant' and a 'nonconstant' part.

Remark 8.5.5. Let $f : X \to Y$ be a projective morphism between smooth varieties that is flat of relative dimension one. The technique used in the case of conic bundles (Example 8.5.1) and in the proof of Theorem 8.5.3 produces a relative Chow-Künneth decomposition for f in the weak sense. M. Saito has proved that a morphism f as above admits a relative Chow-Künneth decomposition in the strong sense if f is of relative dimension at most one; see [**MüS-Sa**, (1.5)].

CHAPTER 9

Beyond Pure Motives

Up to now we have discussed motives for smooth projective varieties over a field k. There is no theory yet for mixed motives, i.e. a theory of motives for arbitrary varieties over a fixed field k, but there are several steps toward such a theory as we explain now.

Serre [**Serre91**, §8] suggested the following approach to "virtual" motives. The idea is that any variety can be obtained from the smooth projective ones by successive cutting and pasting of smooth subvarieties. Encoding this K-theoretically and applying the functor h should then give the "virtual" mixed motives. More precisely, let X be any k-variety and write it as a disjoint union of smooth quasi-projective varieties Y. Now, if resolution of singularities would hold, one may write each Y as the complement of a union of a smooth projective variety and some smaller dimensional variety. By induction one can then write

$$X = \coprod_j Y_j - \coprod_k Z_k$$

where the Y_j and Z_k are smooth projective. The virtual motive should be

$$h_\sim(X) := \sum_j h_\sim(Y_j) - \sum_k h_\sim(Z_k)$$

where this has to be interpreted in an appropriate Grothendieck group, the group $K_0\mathrm{Mot}_\sim(k)$. This K-theoretic approach can be made to work as was shown by Gillet and Soulé in [**Gil-So**], and independently by Guillén and Navarro Aznar [**Gu-Na**]. We follow in § 9.1 a simpler approach which is due to Looijenga and Bittner [**Bitt**].

The approach in [**Gil-So**] and [**Gu-Na**] is richer since it produces a lifting of the "virtual motive" to the homotopy category. This lifting of $h_\sim(X)$ is the weight complex $W(X)$ as explained in § 9.2. The name comes from the Deligne weight complex in mixed Hodge theory. But the motivic weight complex contains more information, since it works over the integers and thus yields a weight filtration on *integral* cohomology groups. Of course this could a priori just be a trivial enhancement of Deligne's weights on rational cohomology, but as shown by an example (cf. § 9.2.3), it is not!

The third and last approach is the one of Voevodsky, who constructed the *triangulated category of motives* [**Maz-Vo-We, Voe00**]. The category of Chow motives embeds in the Voevodsky category provided we reverse arrows. A brief discussion of Voevodsky's theory is given in § 9.3. Results similar to Voevodsky's have been obtained by Hanamura [**Hana95, Hana04**] and Levine [**Lev98**]. The hope is that these triangulated categories of motives carry a suitable t-structure whose heart

123

would give the category of mixed motives. An entirely different approach has been suggested by Nori. Although Nori did not publish his results, a short introduction to his ideas can be found in [**Hub-MüS**] and [**Lev05**].

Two Auxilliary Notions

Let k be any algebraically closed field. We say that k *admits resolution of singularities* if any scheme of finite type over k admits resolution of singularities. By Hironaka's fundamental theorem, algebraically closed fields of characteristic zero admit resolution of singularities.

We say that *weak factorization holds for k* if the following statement holds for every birational map $\phi : X_1 \dashrightarrow X_2$ between complete smooth connected varieties over k:

PROPERTY (Weak Factorization). Let $U \subset X_1$ be an open set where ϕ is an isomorphism. Then ϕ can be factored into a sequence of blowing-ups and blowing-downs with smooth centers disjoint from U, i.e. there exists a diagram

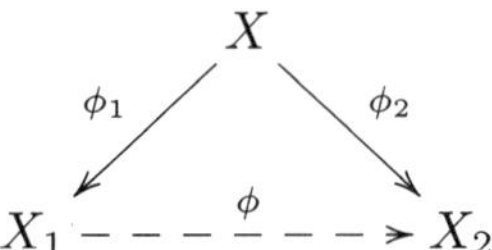

where ϕ_1 and ϕ_2 are projective morphisms which are compositions of blowings up in smooth centers disjoint from (the inverse images of) U. Moreover, if $X_1 - U$ (respectively, $X_2 - U$) is a simple normal crossings divisor, then the factorization can be chosen such that the inverse images of this divisor under the successive blowings up are also simple normal crossing divisors.

By [**Ab-K-M-W**] weak factorization holds for any field of characteristic zero, algebraically closed or not.

9.1. Motivic Euler Characteristics

DEFINITION 9.1.1. Let $\mathsf{Var}(k)$ be the category of k-varieties, i.e. reduced and separated schemes of finite type over k. Product of varieties endows $\mathbb{Z}[\mathsf{Var}(k)]$ with a ring structure.

Define $K_0\mathsf{Var}(k)$ to be the quotient ring $\mathbb{Z}[\mathsf{Var}(k)]/J$ of finite formal sums $\sum n_V(V)$, $n_V \in \mathbb{Z}$, of isomorphism classes (V) of varieties V modulo the ideal J generated by $(X) - (Y) - (X - Y)$ for any closed subvariety $Y \subset X$. Let $[X]$ be the class of X in $K_0\mathsf{Var}(k)$. By construction in $K_0\mathsf{Var}(k))$ the *scissor-relation* $[X] = [Y] + [X - Y]$ holds.

The following result gives a presentation of the Grothendieck group of varieties in terms of generators and relations.

THEOREM 9.1.2 ([**Bitt**, Thm. 3.1]). *Suppose that weak factorization holds for k. Then $K_0\mathsf{Var}(k)$ is generated by classes $[X]$ of smooth projective varieties subject to the following two relations*

(1) $[\varnothing] = 0$.

(2) *the blow-up relation* $[X] - [Z] = [Y] - [E]$ *where* $Y = \mathrm{Bl}_Z(X)$ *is a blow up in a smooth subvariety* $Z \subset X$ *and* E *the exceptional divisor. In other words,*

$$K_0(\mathsf{Var}(k)) = \mathbb{Z}[\mathsf{SmProj}(k)]/I,$$

where I *is the ideal generated by the relations* $[\varnothing] - 0$ *and the blow-up relations* $[X] - [Z] - [Y] + [E]$, *where the smooth projective varieties* X, Y, Z, E *fit in the* blow-up diagram

$$
\begin{array}{ccc}
E & \xrightarrow{\tilde{i}} & Y = \mathrm{Bl}_Z X \\
{\scriptstyle \sigma|E} \downarrow & & \downarrow {\scriptstyle \sigma} \\
Z & \xrightarrow{i} & X.
\end{array}
\tag{62}
$$

Let us come back to the category $\mathsf{Mot}_\sim(k)$ of motives. Recall from § 2.3 that there is a functorial way to assign to a smooth projective variety X a motive $h_\sim(X) \in \mathsf{Mot}_\sim(k)$.

As for any additive (small) category, we can associate a K_0-group to $\mathsf{Mot}_\sim(k)$; it is the free abelian group on the isomorphism classes $[M]$ of motives M modulo the relation $[M] = [M'] + [M'']$ whenever $M \simeq M' \oplus M''$.

PROPOSITION 9.1.3. *The functor h passes to K-theory, i.e., we have a well-defined ring homomorphism, the* motivic Euler characteristic with compact support

$$\chi^c_{\mathrm{mot}} : \quad K_0(\mathsf{Var}(k)) \quad \to \quad K_0\mathsf{Mot}_\sim(k) \tag{63}$$

Proof: The functor

$$h_\sim : \mathsf{SmProj}(k) \to \mathsf{Mot}_\sim(k)$$

induces a ring homomorphism $K_0(\mathsf{SmProj}(k)) \to K_0(\mathsf{Mot}_\sim(k))$ sending $[X]$ to $[h_\sim(X)]$. By Bittner's theorem 9.1.2 this ring homomorphism descends to $K_0(\mathsf{Var}(k))$ if it vanishes on the ideal $I \subset K_0(\mathsf{SmProj}(k))$ generated by the blow-up relations, i.e., if we have

$$h_\sim(X) - h_\sim(Z) = h_\sim(Y) - h_\sim(E)$$

for every blow-up diagram. This relation holds by Manin's computation of the motive of a blow-up; see ((17)). $\qquad\square$

Remark. By construction the ring homomorphism χ^c_{mot} sends the class $[\mathbb{A}^1]$ of the affine line to the Lefschetz motive $\mathbb{L}$.

To define the motivic Euler characteristic (without compact support)

$$\chi_{\mathrm{mot}} : K_0(\mathsf{Var}(k)) \to K_0(\mathsf{Mot}_\sim(k))$$

we shall define a motivic Euler characteristic for pairs. The idea guiding this construction is that the motivic Euler characteristic with compact support of a possibly non-complete variety X should be the same as the motivic Euler characteristic of the pair $(\bar{X}, \partial X)$, where $\bar{X}$ is some completion of X obtained by adding a "boundary" $\partial X := \bar{X} - X$ to X. Once this has been done, the motivic Euler characteristic without support $\chi_{\mathrm{mot}}(X)$ is defined as the motivic Euler characteristic of the pair $(X, \varnothing)$.

To make this precise, one starts from the category $\mathsf{Var}^2(k)$ of pairs (X,Y) of varieties where $Y \subset X$ is a closed subvariety of a variety X. One introduces the corresponding Grothendieck group:

DEFINITION 9.1.4. The Grothendieck group $K_0\mathsf{Var}^2(k)$ is the quotient $\mathbb{Z}[\mathsf{Var}^2(k)]/\tilde{J}$ consisting of finite formal sums $\sum n_{X,Y}(X,Y)$, $n_{X,Y} \in \mathbb{Z}$ of isomorphism classes of pairs $(X,Y) \in \mathsf{Var}^2(k)$ modulo the ideal $\tilde{J}$ generated by

- *Excision*: $(X', f^{-1}Y)) - (X,Y) \in \tilde{J}$ whenever $f : X' \to X$ is proper and $Y \subset X$ is a closed subvariety such that f induces an isomorphism $X' - f^{-1}Y \cong X - Y$;
- *Gysin maps*: $(X - D, \varnothing) - [(X, \varnothing) - (\mathbb{P}^1 \times D, \{\infty\} \times D)] \in \tilde{J}$ whenever X is smooth and connected, and $D \subset X$ is a smooth divisor;
- *Exactness*: $(X, Z) - (X, Y) - (Y, Z) \in \tilde{J}$ whenever $X \supset Y \supset Z$, Y closed in X and Z closed in Y.

Denote the isomorphism class of (X,Y) in this group by $\{X,Y\}$. The class of $(X, \varnothing)$ is also denoted $\{X\}$. One can further show that the product operation

$$\{X,Y\} \cdot \{X',Y'\} := \{X \times X', X \times Y' \cup Y \times X'\}$$

respects excision, exactness (both easy), and the Gysin relation (less trivial, see [**Bitt**, § 4]). Consequently, $K_0\mathsf{Var}^2(k)$ becomes a ring. Note that excision implies that $\{X,Y\}$ only depends on the isomorphism class of the pair (X,Y). By exactness $\{X,Y\} = \{X\} - \{Y\}$ while exactness and excision yield $\{X \sqcup Y\} = \{X \sqcup Y, Y\} + \{Y\} = \{X\} + \{Y\}$ and $\{\varnothing\} = 0$. If $X \subset W$ is an open embedding with W complete, then the the excision property implies that $\{W, W - X\}$ is independent of W. We set

$$\{X\}_c := \{W, W - X\}.$$

If $Y \subset X$ is closed this can be seen to imply

$$\{X\}_c = \{X - Y\}_c + \{Y\}_c.$$

Hence we have a ring homomorphism

$$\psi : K_0(\mathsf{Var}(k)) \to K_0(\mathsf{Var}^2(k)), \quad \{X\} \mapsto \{X\}_c. \tag{64}$$

THEOREM 9.1.5 ([**Bitt**, Theorem 4.2]). *Suppose that weak factorization holds for k. Then the above ring homomorphism is an isomorphism.*

Using Theorem 9.1.5 we can define the *motivic Euler characteristic for pairs* as

$$\chi^2_{\mathrm{mot}} = \chi^c_{\mathrm{mot}} \circ \psi^{-1} : K_0\mathsf{Var}^2(k) \quad \to \quad K_0\mathsf{Mot}_\sim(k)$$

In other words, χ^2_{mot} is the unique map that makes the following diagram commutative:

$$
\begin{array}{ccc}
K_0\mathsf{Var}(k) & \xrightarrow{\ \chi^c_{\mathrm{mot}}\ } & \\
\ \ \downarrow{\scriptstyle \psi} & \xrightarrow[\ \chi^2_{\mathrm{mot}}\ ]{} & K_0\mathsf{Mot}_\sim(k) \\
K_0\mathsf{Var}^2(k). & &
\end{array}
$$

We then define the *motivic Euler characteristic* as the ring homomorphism

$$\chi_{\mathrm{mot}} : K_0(\mathsf{Var}(k)) \to K_0(\mathsf{Mot}_\sim(k))$$

defined by $\chi_{\mathrm{mot}}[X] = \chi^2_{\mathrm{mot}}(\{X, \varnothing\})$.

Remark. If we invert the class $[\mathbf{A}^1]$ of the affine line, we obtain the so-called *naive motivic ring*

$$\mathcal{M}(k) := K_0\mathsf{Var}(k)[\mathbf{A}^1]^{-1}.$$

This ring carries a *duality involution* D that is characterized by the property

$$\mathsf{D}[X] = [\mathbf{A}^1]^{-d}[X], \quad X \text{ smooth and projective}, d = \dim X.$$

The two Euler characteristics χ^c_{mot} and χ_{mot} are interchanged by the duality operator:

$$\mathsf{D}[\chi_{\mathrm{mot}}\{X\}] = \mathbf{L}^{-d} \otimes \chi^c_{\mathrm{mot}}[X] = \chi^c_{\mathrm{mot}}\mathsf{D}([X]).$$

9.2. Motives Via Weights

9.2.1. Weight Complex, Statement of Main Result.
In the previous section, we have defined motivic Euler characteristics with values in $K_0(\mathsf{Mot}_\sim(k))$. We want to lift these to have values in $\mathbf{H}^b(\mathsf{Mot}_\sim(k))$, where for any additive category $\mathfrak{A}$ the notation $\mathbf{H}^b(\mathfrak{A})$ stands for the category of *bounded cochain complexes in $\mathfrak{A}$ up to homotopy.*

Let us briefly review this notion. To start, a *homotopy operator* between two morphisms $f, g : K \to L$ between complexes K and L in an additive category $\mathfrak{A}$ is a collection of morphisms $k^q : K^q \to L^{q-1}$ such that $f^q - g^q = d^{q-1} \circ k^q + k^{q+1} \circ d^q$. This is an equivalence relation compatible with composition. If such a homotopy-operator exists we write $f \sim g$ and we say that f is *homotopic* to f. The objects in the homotopy category $\mathbf{H}^b(\mathfrak{A})$ are the bounded complexes in $\mathfrak{A}$ and the morphisms are classes of homotopic cochain maps. Isomorphic objects in this category are those which are *homotopy equivalent*, i.e admiting a *homotopy equivalence* between them: a morphism admitting an inverse up to homotopy.

As a first step toward the weight complex, it can be shown that

$$K_0(\mathbf{H}^b(\mathsf{Mot}_\sim(k)) = K_0(\mathsf{Mot}_\sim(k));$$

see e.g. [**Gil-So**, Lemma 3].

THEOREM 9.2.1 ([**Gil-So**, Thm. 2] [**Gu-Na**, Th. 5.2]). *For every $X \in \mathsf{Var}(k)$ there is a unique complex (up to homotopy) $W(X) \in \mathbf{H}^b\mathsf{Mot}_\sim(k)$, called the* weight complex. *It refines the motivic Euler characteristic in the sense that $\chi^c_{\mathrm{mot}}(X) = [W(X)] \in K_0(\mathbf{H}^b(\mathsf{Mot}_\sim(k)) = K_0(\mathsf{Mot}_\sim(k))$.*
The assigment $X \mapsto W(X)$ is functorial:

- *a proper map $f : X \to Y$ induces $f^* : W(Y) \to W(X)$ and composable proper maps f, g satisfy $(g \circ f)^* = f^* \circ g^*$; in other words, if $\mathsf{Var}_c(k)$ denotes the category of k-varieties with proper morphisms between them, we get a contravariant functor*

$$W : \mathsf{Var}_c(k) \to \mathbf{H}^b\mathsf{Mot}_\sim(k).$$

- *open immersions $i : U \hookrightarrow X$ induce $i_* : W(U) \to W(X)$; composable open immersion behave likewise functorial;*

It obeys the product rule $W(X \times Y) = W(X) \otimes W(Y)$ and it has a strong *motivic property refining the scissor-relations: If $i : Y \hookrightarrow X$ is a closed immersion with complement $j : U = X - Y \hookrightarrow X$ one has a distinguished triangle*

$$W(U) \xrightarrow{j_*} W(X) \xrightarrow{i^*} W(Y) \to W(U)[1].$$

Remark. Guillén and Navarro Aznar have defined a contravariant functor

$$G : \mathsf{Var}(k) \to \mathbf{H}^b\mathsf{Mot}_\sim(k),$$

the Gysin complex that refines the motivic Euler characteristic $\chi_{\mathrm{mot}} : K_0(\mathsf{Var}(k)) \to K_0(\mathsf{Mot}_\sim(k))$ in the sense that the class of $G(X)$ in $K_0(\mathsf{Mot}_\sim(k))$ equals $\chi_{\mathrm{mot}}(X)$. See [**Gu-Na**, § 5.6–5.10].

9.2.2. On the Construction of the Weight Complex. The construction of the weight complex can be understood as being based on the following two principles.

(1) The weight complex reflects the deviation of purity: in the case of a smooth projective variety X the weight complex is the motive $h_\sim(X)$ considered as a homotopy class of a complex concentrated in degree 0.

(2) The cutting and pasting procedure using blow-ups must be replaced by homotopy.

To explain the second principle, we start by recalling that the category of Chow-motives $\mathsf{Mot}_{\mathrm{rat}}(k)$ contains the motives $\mathsf{ch}(X) = h_{\mathrm{rat}}(X)$ of smooth projective k-varieties as well as their Tate twists $\mathsf{ch}(X)(j)$ and arbitrary finite sums of such motives. If X and Y are two smooth projective k-varieties and Γ a degree zero correspondence from Y to X, there is an induced morphism $\mathsf{ch}(\Gamma) : \mathsf{ch}(Y) \to \mathsf{ch}(X)$. In particular, if $f : X \to Y$ is a morphism ofvarieties the transpose of the graph defines $\mathsf{ch}(f) : \mathsf{ch}(Y) \to \mathsf{ch}(X)$. We can view this as a morphism of cochain complexes in $\mathsf{Mot}_{\mathrm{rat}}(k)$ concentrated in degree 0. Then its cone, $\mathrm{Cone}(\mathsf{ch}(f))$ (see Definition 8.2.4) is the cochain complex $\mathsf{ch}(Y) \to \mathsf{ch}(X)$, where we place $\mathsf{ch}(X)$ in degree 0 and $\mathsf{ch}(Y)$ in degree -1.

This construction leads to the definition of the weight complex for any k-variety starting with the simplest variety $X - Z$, where $i : Z \to X$ is the inclusion of a smooth subvariety Z of codimension $r + 1$ in a smooth and proper variety X. In this special case we define the *weight complex* of $X - Z$ as the cochain complex $\mathrm{Cone}(\mathsf{ch}(i))[-1]$ in $\mathbf{H}^b\mathsf{Mot}_{\mathrm{rat}}(k)$, i.e.,

$$W(X - Z) = \Big[\underbrace{\mathsf{ch}(X)}_{\deg\,0} \longrightarrow \underbrace{\mathsf{ch}(Z)}_{\deg\,1} \Big], \tag{65}$$

where the rectangular parentheses mean 'isomorphism class in the homotopy category'. To make sense out of this, the weight complex should (up to homotopy equivalence) only depend on $X - Z$ and not on the choice of X and Z.

Let us for instance look at what happens if we blow up Z. Recall the blow-up diagram

$$
\begin{array}{ccc}
E & \xrightarrow{\ \tilde{\imath}\ } & Y = \mathrm{Bl}_Z X \\
{\scriptstyle \sigma|E}\downarrow & & \downarrow{\scriptstyle \sigma} \\
Z & \xrightarrow{\ i\ } & X.
\end{array}
$$

We should then prove that $W(X - Z) = W(Y - E)$:

LEMMA 9.2.2. *In* $\mathsf{Mot}_{\mathrm{rat}}(k)$ *the morphism of cochain complexes*

$$\mathrm{Cone}(i) \to \mathrm{Cone}(\tilde{\imath})$$

induced by the pair $(\mathsf{ch}(\sigma|E), \mathsf{ch}(\sigma))$ *is a homotopy equivalence. Consequently, in* $\mathsf{H}^b(\mathsf{Mot}_{\mathrm{rat}}(k))$ *one has equality* $W(X - Z) = W(Y - E)$.

Proof: Manin's calculations (see (17)) on the motive of a blown up variety show that there are canonical isomorphisms of motives

$$\mathsf{ch}(E) \;=\; \mathsf{ch}(Z) \oplus T, \qquad T := \bigoplus_{j=1}^{r} \mathsf{ch}(Z)(-j)$$
$$\mathsf{ch}(Y) \;=\; \mathsf{ch}(X) \oplus T$$

induced by degree 0 correspondences between the left and right hand sides of the two formulas. In fact, from Manin's arguments [**Manin**], it follows that in the category $\mathsf{Mot}_{\mathrm{rat}}(k)$ the blow-up diagram (62) is canonically isomorphic to

$$
\begin{array}{ccc}
\mathsf{ch}(Z) \oplus T & \xleftarrow{\;\;\mathbf{i}:=\begin{pmatrix} \mathsf{ch}(i) & 0 \\ 0 & \mathrm{id}_T \end{pmatrix}\;\;} & \mathsf{ch}(X) \oplus T \\[4pt]
{\scriptstyle i'_1 = \mathsf{ch}(\sigma|E)}\Big\uparrow & & {\scriptstyle i_1 = \mathsf{ch}(\sigma)}\Big\uparrow \\[4pt]
\mathsf{ch}(Z) & \xleftarrow{\;\;\mathsf{ch}(i)\;\;} & \mathsf{ch}(X),
\end{array}
\tag{66}
$$

where i_1 and i'_1 are the inclusions of the first factors. Let p_1 and p'_1 be the corresponding projections.

To show that (i'_1, i_1) is a homotopy equivalence it suffices to show that the four compositions $p_1 \circ i_1, p'_1 \circ i'_1, i_1 \circ p_1$ and $i'_1 \circ p'_1$ are homotopic to the identity. Clearly, $p_1 \circ i_1 = \mathrm{id}$, $p'_1 \circ i'_1 = \mathrm{id}$. For the remaining two compositions, introduce the map

$$\mathbf{k} = \begin{pmatrix} 0 & 0 \\ 0 & \mathrm{id}_T \end{pmatrix} : \mathsf{ch}(X) \oplus T \to \mathsf{ch}(Z) \oplus T.$$

It serves as a homotopy operator between (i'_1, i_1) and the identity: since

$$i_1 \circ p_1 = \begin{pmatrix} \mathrm{id}_X & 0 \\ 0 & 0 \end{pmatrix} \;=\; \begin{pmatrix} \mathrm{id}_X & 0 \\ 0 & \mathrm{id}_T \end{pmatrix} - \mathbf{i} \circ \mathbf{k}$$
$$i'_1 \circ p'_1 = \begin{pmatrix} \mathrm{id}_Z & 0 \\ 0 & 0 \end{pmatrix} \;=\; \begin{pmatrix} \mathrm{id}_Z & 0 \\ 0 & \mathrm{id}_T \end{pmatrix} - \mathbf{k} \circ \mathbf{i}$$

showing that these two compositions are indeed homotopic to the identity. $\qquad\square$

The above principles can be refined to give a definition of the weight complex in general which is independent of choices. For details we refer to [**Gil-So**]. We illustrate the definition with an example.

EXAMPLE 9.2.3 ([**Gil-So**, Theorem 3]). Let $X \supset U$ be a *good compactification* of U. This means that X itself is a smooth projective variety and that $D := X - U = \bigcup_{i \in I} D_i$ is a union of smooth hypersurfaces D_i of X meeting like the coordinate hyperplanes in $\mathbf{A}^d$. Set

$$
\begin{aligned}
D_I &= D_{i_1} \cap \cdots \cap D_{i_m}, \quad I = \{i_1, \dots, i_m\} \\
a_I &: D_I \hookrightarrow X \\
D(m) &= \coprod_{|I|=m} D_I \\
a_m &= \coprod_{|I|=m} a_I : D(m) \hookrightarrow X.
\end{aligned}
$$

The $D(m)$ are smooth projective. By definition, $D_\varnothing = X$. For $I = (i_1, \dots, i_m)$ put

$$I_j = (i_1, \dots, \widehat{i_j}, \dots, i_m).$$

There are m natural inclusions $a_I^j : D_I \hookrightarrow D_{I_j} \subset D(m-1)$ which can be assembled into morphisms

$$\delta_m : D(m) \to D(m-1), \quad \delta_m|D_I = \sum_{j=1}^{m}(-1)^j a_I^j$$

and which satisfy $\delta_{m-1}\circ\delta_m = 0$. So this gives a chain complex concentrated in degrees $k, k-1, \ldots, 0$ where k is the maximal m for which $D(m)$ is not empty. Applying the functor h and the appropriate shift, we get the weight complex

$$W(U) := \left[0 \to \underbrace{\mathsf{ch}(X)}_{0} \to \underbrace{\mathsf{ch}(D(1))}_{1} \to \cdots \to \underbrace{\mathsf{ch}(D(k))}_{k} \to 0 \right].$$

This complex is related to $H_c^*(U) = H^*(X, D)$ via the so-called *weight spectral sequence* for compactly supported cohomology whose E_2-term is given by

$$E_2^{j,n} = H^j\left(\cdots H^n(D(j-1)) \xrightarrow{\delta_j^*} H^n(D(j)) \xrightarrow{\delta_{j+1}^*} H^n(D(j+1))\cdots \right)$$

and which converges to $H^*(X, D)$:

$$E_2^{j,n} \implies H^{j+n}(X, D) = H_c^{j+n}(U). \tag{67}$$

This spectral sequence refines the weight spectral sequences in mixed Hodge theory with $\mathbb{Q}$-coefficients, which in this case degenerates at E_2.

EXAMPLE 9.2.4. We come back to the case where U can be compactified to a smooth projective X with a single smooth divisor D. In this case $H_c^n(U)$ carries a two-step weight filtration with graded pieces

$$\begin{aligned}
\mathrm{Gr}_{n-1}^W H_c^n(U) &= E_2^{1,n-1} = \mathrm{Coker}(i^* : H^{n-1}(X) \to H^{n-1}(D)) \\
\mathrm{Gr}_n^W H_c^n(U) &= E_2^{0,n} = \mathrm{Ker}(i^* : H^n(X) \to H^n(D))
\end{aligned}$$

as it should be in mixed Hodge theory.

9.2.3. Weight Filtration on Integral Cohomology. Suppose we are given a covariant functor from motives to some abelian category $\mathfrak{A}$, say $\Gamma : \mathsf{Mot}_\sim(k) \to \mathfrak{A}$. For any variety X this yields a complex $\Gamma(W(X))$ in $\mathfrak{A}$ for which one may calculate the cohomology $H^i\Gamma(W(X))$. Theorem 9.2.1 implies for instance that this assigment is contravariantly functorial in X and one has long exact sequences associated to pairs (X, Y) where $Y \subset X$ is a closed subvariety.

Note that to define $W(X)$ we work with complexes with *integer* coefficients. So we may take for Γ the functor H^n, integral cohomology of rank n. Applying (67) then yields a weight filtration on *integral* cohomology with compact support which induces the one on rational cohomology from Hodge theory constructed in [**Del71, Del74a**]. However, this weight filtration cannot in any way be deduced from Deligne's weight filtration and is a truly motivic invariant as illustrated by the following example.

EXAMPLE 9.2.5 (compare [**Gil-So**, p. 148]). Let S be a Kummer surface (i.e. the *singular* quotient of an abelian surface A by the standard involution ι) and let S' an Enriques surface. We claim that $H^3(S \times S') = (\mathbb{Z}/2\mathbb{Z})^6$ with 2-step weight filtration $H^3 = W_3 \supset W_2$, $W_2 = (\mathbb{Z}/2\mathbb{Z})^5$.

To see this, we first calculate the cohomology of S by comparing it to that of the minimal resolution $\sigma : \tilde{S} \to S$ which is a K3-surface. We know [**Bar-Hu-Pe-VdV**,

Ch. VIII §3] that $H^1(\tilde{S}) = H^3(\tilde{S}) = 0$ while $H^2(\tilde{S}) = \mathbb{Z}^{22}$. Let $E_i \subset \tilde{S}$, $i = 1,\ldots,16$ be the exceptional curves and consider the Mayer-Vietoris sequence for the pair $(\tilde{S}, \cup E_i)$. Let $j : \cup E_i \hookrightarrow S$ be the inclusion. Then the Mayer-Vietoris sequence has a piece

$$0 \to H^2(S) \xrightarrow{\sigma^*} H^2(\tilde{S}) \xrightarrow{j^*} \oplus_i H^2(E_i) \to H^3(S) \to 0. \qquad (68)$$

Since $S = A/\iota$, and ι acts as the identity on $H^2(A)$ we have $H^2(S) = H^2(A) = \mathbb{Z}^6$ and $H^3(S)$ must be torsion. The homology sequence is dual to the above and reads

$$0 \to \oplus_i H_2(E_i) \xrightarrow{j_*} H_2(\tilde{S}) \xrightarrow{\sigma_*} H_2(S) \to 0.$$

By [loc. cit. Ch. VIII § 5], the lattice $\Gamma_0 = \mathrm{Im}(j_*) = \mathrm{Ker}(\sigma_*)$ spanned by the fundamental classes of the E_i is a sublattice of the primitive lattice Γ it generates inside $H^2(\tilde{S})$ of index 2^5. Moreover, Γ/Γ_0 is 2-torsion so that

$$\mathrm{Tor}(H_2(S)) = \Gamma/\Gamma_0 \simeq (\mathbb{Z}/2\mathbb{Z})^5.$$

The first equality follows since σ_* maps Γ/Γ_0 isomorphically to $\mathrm{Tor}(H_2(S))$. By the universal coefficient theorem one thus has that

$$\mathrm{Tor}(H_2(S)) \simeq \mathrm{Tor}(H^3(S)) = H^3(S) \simeq (\mathbb{Z}/2\mathbb{Z})^5.$$

The weight complex is compatible with the Mayer-Vietoris sequence because of the motivic property (Theorem 9.2.1) so the last map of the exact sequence (68) is a surjection of weight 2 spaces and hence $H^3(S)$ has pure weight 2.

One also knows the invariants of an Enriques surface [loc. cit, Ch. VIII,§ 15]. For instance $H^1(S') = 0$, $H^3(S') = \mathbb{Z}/2\mathbb{Z}$ while $H^2(S') = \mathbb{Z}^{10} \oplus \mathbb{Z}/2\mathbb{Z}$. Again by functoriality the weight filtration on the k-th cohomology of smooth varieties must be pure of weight k and hence $H^3(S')$ has pure weight 3.

The Mayer-Vietoris sequence shows that $H^1(S) = 0$. Hence the Künneth formula simplifies to give $H^3(S \times S') = H^3(S) \oplus H^3(S')$. The product rule for the weight filtration coupled to functoriality implies that the Künneth formula respects weight. It follows that $H^3(S \times S')$ indeed has a non-trivial two-step weight filtration $W_3 \supset W_2$.

9.3. Voevodsky's Triangulated Category of Motives

Throughout this section we assume that k is a perfect field that admits resolution of singularities.

We discuss (without proofs) the construction of Voevodsky's category $\mathsf{DM}^{\mathrm{eff}}_{\mathrm{gm}}(k)$ of *effective geometric motives*, its basic properties and its relation to the category $\mathsf{Mot}_{\mathrm{rat}}(k)$ of Chow motives. For a thorough discussion of Voevodsky's theory see [**Maz-Vo-We**]; see also [**Andr**],[**Friedl-Su-Vo**] and [**Lev98**].

9.3.1. Effective Geometric Motives.

DEFINITION 9.3.1. Given X, $Y \in \mathsf{Var}(k)$, the group $\mathsf{Corr}_{\mathrm{fin}}(X,Y)$ of *finite correspondences* from X to Y is the abelian group generated by integral subschemes $Z \subset X \times Y$ such that $p_X : Z \to X$ is finite and Z surjects onto an irreducible component of X.

The category $\mathsf{SmCor}(k)$ is the category with objects $[X]$, $X \in \mathsf{Sm}(k)$ and morphisms $\mathrm{Hom}(X,Y) = \mathsf{Corr}_{\mathrm{fin}}(X,Y)$. The graph Γ_f of a morphism $f : X \to Y$ is a finite correspondence from X to Y, which we shall denote by f_*. There exists a functor $\mathsf{Sm}(k) \to \mathsf{SmCor}(k)$ that sends X to $[X]$ and $f : X \to Y$ to $\Gamma_f = f_* \in \mathsf{Corr}_{\mathrm{fin}}(X, Y)$.

The category $\mathsf{SmCor}(k)$ is an additive tensor category with tensor product

$$[X] \otimes [Y] := [X \times Y].$$

The bounded homotopy category $\mathbf{H}^b(\mathsf{SmCor}(k))$ is a triangulated tensor category.

DEFINITION 9.3.2. The category $\mathsf{DM}^{\mathrm{eff}}_{\mathrm{gm}}(k)$ of *effective geometric motives* is obtained in the following way:

(1) Localize[1] $\mathbf{H}^b(\mathsf{SmCor}(k))$ with respect to the thick subcategory generated by complexes of the form
  (a) [Homotopy] $[X \times \mathbf{A}^1] \to [X]$;
  (b) [Mayer-Vietoris] $[U \cap V] \to [U] \oplus [V] \to [X]$, where U and V are Zariski open subsets of X such that $X = U \cup V$.
(2) Take the pseudo–abelian completion of the resulting quotient category.

The category $\mathsf{DM}^{\mathrm{eff}}_{\mathrm{gm}}(k)$ then is a triangulated tensor category [**Bal-Sc**].

DEFINITION 9.3.3. The *motive* of $X \in \mathsf{Sm}(k)$ (in the sense of Voevodsky) is the image of $[X]$ in $\mathsf{DM}^{\mathrm{eff}}_{\mathrm{gm}}(k)$. We denote this motive by $M(X)$.

The covariant functor $M : \mathsf{Sm}(k) \to \mathsf{DM}^{\mathrm{eff}}_{\mathrm{gm}}(k)$ sends $f : X \to Y$ to $f_* : M(X) \to M(Y)$ and satisfies

$$M(X \textstyle\coprod Y) = M(X) \oplus M(Y), \quad M(X \times Y) = M(X) \otimes M(Y). \tag{69}$$

Remark. Note that Voevodsky's construction is *covariant* rather than contravariant, i.e., it gives *homological motives* rather than cohomological motives.

Remark. For the definition of $\mathsf{DM}^{\mathrm{eff}}_{\mathrm{gm}}(k)$ we have followed [**Friedl-Su-Vo**, Chapter 5]. In [**Maz-Vo-We**] this category is defined in a different way. Both definitions are equivalent; see [loc.cit., p. 110].

9.3.2. Properties.

We list the main properties of the category $\mathsf{DM}^{\mathrm{eff}}_{\mathrm{gm}}(k)$. The properties M1-M3 readily follow from the definition. The finer properties M4-M6 are proved via an embedding of $\mathsf{DM}^{\mathrm{eff}}_{\mathrm{gm}}(k)$ into a triangulated category $\mathsf{DM}^{\mathrm{eff}}_-(k)$, the category of *motivic complexes*, whose construction will be sketched in Appendix E.

Basic Properties.
M1 (Homotopy-invariance) $M(X \times \mathbf{A}^1) \simeq M(X)$.
M2 (Mayer-Vietoris triangle)

$$M(U \cap V) \to M(U) \oplus M(V) \to M(X) \to M(U \cap V)[1].$$

M3 (Tate object) Let $a : \mathbb{P}^1_k \to \mathrm{Spec}\, k$ be the structure morphism. The *Tate object* $\mathbb{Z}(1)$ is the complex

$$[\mathbb{P}^1] \xrightarrow{a_*} [\mathrm{Spec}\, k]$$

cf. [**Friedl-Su-Vo**, p. 192]. We write $\mathbb{Z}(n) = \mathbb{Z}(1)^{\otimes n}$. We denote by $\mathsf{DM}_{\mathrm{gm}}(k)$ the triangulated category that is obtained from $\mathsf{DM}^{\mathrm{eff}}_{\mathrm{gm}}(k)$ by inverting the motive $\mathbb{Z}(1)$.

---

[1]see e.g. [**Weib**, Thm. 10.3]

Voevodsky has shown that one can associate to *every* $X \in \mathsf{Var}(k)$ an effective geometric motive $M(X)$; cf. Appendix E. This extends the construction we have seen in § 9.3.1 for smooth varieties. Voevodsky [**Voe00**] shows that the motive $M(X)$ satisfies the following properties.

M4 (Blow-up triangle) Every blow-up diagram

$$
\begin{array}{ccc}
E & \xrightarrow{\tilde{\imath}} & Y = \mathrm{Bl}_Z X \\
{\scriptstyle \sigma|E}\downarrow & & \downarrow{\scriptstyle \sigma} \\
Z & \xrightarrow{i} & X.
\end{array}
$$

induces a distinguished triangle

$$
M(E) \xrightarrow{(\sigma|E)_* + \tilde{\imath}_*} M(Z) \oplus M(Y) \xrightarrow{i_* - \sigma_*} M(X) \to M(E)[1] \tag{70}
$$

in $\mathsf{DM}^{\mathrm{eff}}_{\mathrm{gm}}(k)$.

M5 (Gysin triangle) If X is a smooth scheme, and $Z \subset X$ is a smooth closed subscheme of codimension c in X, there exists a distinguished triangle

$$
M(X - Z) \to M(X) \to M(Z)(c)[2c] \to M(X - Z)[1].
$$

M6 (Duality involution) There exists a duality involution $\mathsf{D} : \mathsf{DM}_{\mathrm{gm}}(k) \to \mathsf{DM}_{\mathrm{gm}}(k)$ such that

$$
\mathrm{Hom}_{\mathsf{DM}_{\mathrm{gm}}}(M(X), M(Y)) \cong \mathrm{Hom}_{\mathsf{DM}_{\mathrm{gm}}}(M(X) \otimes \mathsf{D}(M(Y)), \mathbb{Z}) \tag{M6a}.
$$

If X is smooth and proper of dimension d, then

$$
\mathsf{D}(M(X)) \simeq M(X)(-d)[-2d] \tag{M6b}.
$$

Remark. There exists a second motive, the *motive with compact support* $M^c(X) \in \mathsf{DM}^{\mathrm{eff}}_{\mathrm{gm}}(k)$; see Appendix E. If X is proper, then $M^c(X) = M(X)$.

Motivic cohomology. Recall that if X is defined over $\mathbb{C}$ and

$$
a_X : X \to \operatorname{Spec} \mathbb{C}
$$

is the structure morphism, the Betti cohomology $H^p(X(\mathbb{C}), \mathbb{Q})$ is given by

$$
\begin{aligned}
H^p(X(\mathbb{C}), \mathbb{Q}) & \cong \ \mathrm{Hom}_{D^b(\mathrm{pt})}(\mathbb{Q}, (a_X)_* \mathbb{Q}_X[p]) \\
& \cong \ \mathrm{Hom}_{D^b(\mathbb{Q}_X)}(\mathbb{Q}_X, \mathbb{Q}_X[p]).
\end{aligned}
$$

The motivic analogue is the following. Voevodsky [**Voe00**] defines:

DEFINITION 9.3.4. Given $X \in \mathsf{Var}(k)$ and $q \geq 0$, the *motivic cohomology groups* of X are the groups

$$
H^p(X, \mathbb{Z}(q)) := \mathrm{Hom}_{\mathsf{DM}}(M(X), \mathbb{Z}(q)[p]). \tag{71}
$$

These turn out to be related to S. Bloch's *higher Chow groups* $\mathsf{CH}^p(X, n)$, which generalize the classical Chow groups and which are defined in the following way. Let

$$
\Delta^n = \operatorname{Spec} k[t_0, \ldots, t_n]/(\sum_{i=0}^{n} t_i - 1)
$$

be the algebraic n-simplex, and let $z^p(X, n)$ be the free abelian group on integral algebraic subvarieties of codimension p in $X \times \Delta^n$ which meet all the faces $X \times \Delta^m$

with $m < n$ properly, i.e., in codimension p. The alternating sum of restriction maps to the faces of Δ^q defines a chain complex

$$\cdots \to z^p(X, n+1) \to z^p(X, n) \to z^p(X, n-1) \to \cdots$$

whose homology groups are the higher Chow groups $\mathsf{CH}^p(X, n)$. These indeed reappear as motivic cohomology:

THEOREM 9.3.5 ([**Maz-Vo-We**, Theorem 19.1]). *If $X \in \mathsf{SmProj}(k)$, then*

$$\mathsf{CH}^p(X, q) \cong H^{2p-q}(X, \mathbb{Z}(p)).$$

9.3.3. Comparison to Chow Motives.

The following result relates Voevodsky's triangulated category of motives to the category of Chow motives.

THEOREM 9.3.6 ([**Maz-Vo-We**, Prop.20.1]). *There exists a fully faithful embedding*

$$i : \mathsf{Mot}_{\mathrm{rat}}(k)^{\mathrm{opp}} \to \mathsf{DM}_{\mathrm{gm}}^{\mathrm{eff}}(k).$$

Proof: Given $X \in \mathsf{SmProj}(k)$, put $i(\mathsf{ch}(X)) = M(X)$. To obtain a well–defined functor, we have to show that degree zero correspondences from Chapter 2 act on Voevodsky motives. This is done as follows. Consider

$$\Gamma \in \mathrm{Hom}_{\mathsf{Mot}_{\mathrm{rat}}(k)^{\mathrm{opp}}}(X, Y) = \mathsf{Corr}_{\mathrm{rat}}^0(Y, X) = \mathsf{CH}^{d_Y}(Y \times X) = \mathsf{CH}_{d_X}(X \times Y).$$

By the moving lemma of Friedlander–Lawson [**Friedl-La**], we can move ${}^{\mathsf{T}}\Gamma \in \mathsf{CH}_{d_X}(X \times Y)$ within its rational equivalence class so that ${}^{\mathsf{T}}\Gamma$ meets the family of cycles $\{x\} \times Y$, $x \in X$, properly. This implies that ${}^{\mathsf{T}}\Gamma \sim_{\mathrm{rat}} \sum_i n_i \Gamma_i$ with $\sum_i n_i \Gamma_i \in \mathsf{Corr}_{\mathrm{fin}}(X, Y)$. The homotopy property $M(X) = M(X \times \mathbf{A}^1)$ implies that two finite correspondences in $\mathsf{Corr}_{\mathrm{fin}}(X, Y)$ that are rationally equivalent induce the same map $M(X) \to M(Y)$. Hence ${}^{\mathsf{T}}\Gamma$ induces a well–defined map ${}^{\mathsf{T}}\Gamma_* : M(X) \to M(Y)$.

We show now that i is fully faithful. So let X and Y both smooth and projective. We use Theorem 9.3.5 and the basic properties listed above to complete the proof as follows:

$$
\begin{aligned}
\mathrm{Hom}_{\mathsf{DM}}(M(X), M(Y)) \quad &\cong_{\mathrm{Property\ (\mathbf{M6a})}} \quad \mathrm{Hom}_{\mathsf{DM}}(M(X) \otimes DM(Y), \mathbb{Z}) \\
&\cong_{\mathrm{Property\ (\mathbf{M6b})}} \quad \mathrm{Hom}_{\mathsf{DM}}(M(X) \otimes M(Y)(-d_Y)[-2d_Y], \mathbb{Z}) \\
&\cong_{\mathrm{Relation\,(69)}} \quad \mathrm{Hom}_{\mathsf{DM}}(M(X \times Y), \mathbb{Z}(d_Y)[2d_Y]) \\
&\cong_{\mathrm{Def.\ (71)}} \quad H^{2d_Y}(X \times Y, \mathbb{Z}(d_Y)) \\
&\cong_{\mathrm{Thm.\ 9.3.5}} \quad \mathsf{CH}^{d_Y}(X \times Y) = \mathsf{Corr}_{\mathrm{rat}}^0(Y, X). \qquad \square
\end{aligned}
$$

Appendix E: The Category of Motivic Complexes

We sketch Voevodsky's construction of the category $\mathsf{DM}^{\mathrm{eff}}_-(k)$ of effective *motivic complexes*, which provides a more flexible sheaf-theoretic framework for dealing with motives. This enables Voevodsky to prove some of the more delicate properties of $\mathsf{DM}^{\mathrm{eff}}_{\mathrm{gm}}(k)$, such as the computation of Hom groups and the existence of blow-up and Gysin triangles.

Nisnevich Sheaves with Transfers. The well known notion of a presheaf F (of sets, groups, ...) on a topological space X can be expressed in categorical terms as a contravariant functor F from the category of open subsets of X to the category of sets, groups, ... As such we can replace the category of open sets of X by any category $\mathfrak{C}$: a presheaf F on $\mathfrak{C}$ with values in a category $\mathfrak{A}$ is simply a contravariant functor $F : \mathfrak{C}^{\mathrm{opp}} \to \mathfrak{A}$.

A *presheaf with transfers* is a presheaf on the category $\mathsf{SmCor}(k)$ with values in the category of abelian groups. Since the objects of $\mathsf{SmCor}(k)$ smooth varieties, such a presheaf F attaches an abelian group $F(X)$ to every smooth variety X. As morphisms from X to Y in $\mathsf{SmCor}(k)$ are finite correspondences Z from X to Y, due to contravariance, every $Z \in \mathsf{Corr}_{\mathrm{fin}}(X, Y)$ gives a homomorphism

$$\mathrm{Tr}(Z) : F(Y) \to F(X),$$

which is the *transfer map* the terminology "presheaf with transfers" refers to. Such a presheaf is called *homotopy invariant* if the natural map

$$p_X^* : F(X) \to F(X \times \mathbf{A}^1)$$

is an isomorphism for all $X \in \mathsf{Sm}(k)$.

DEFINITION. A family of étale morphisms $\{p_i : U_i \to X\}$ satisfies the *Nisnevich lifting property* if for every point $x \in X$ there exist an index i and a point $u \in U_i$ such that the induced map of residue fields $k(x) \to k(u)$ is an isomorphism. If this condition is satisfied, we say that $\{p_i : U_i \to X\}$ is a *Nisnevich covering* of X.

Remark. In the definition of a Nisnevich covering, one requires that the Nisnevich lifting property is satisfied not only for closed points, but for every scheme-theoretic point of X (including the generic point). This implies that if $\{p_i : U_i \to X\}$ is a Nisnevich covering, for every point $x \in X$ there exist a nonempty Zariski open subset $V \subset X$ containing x and an index i such that $p_i^{-1}(V) \to V$ admits a section; cf. [**Maz-Vo-We**, Lemma 12.3].

Nisnevich coverings satisfy the axioms for a Grothendieck pre-topology and generate a Grothendieck topology, the *Nisnevich topology*, on $\mathsf{Var}(k)$. A presheaf with transfers is called a *Nisnevich sheaf with transfers* if its restriction to $\mathsf{Sm}(k)$ is a sheaf for the Nisnevich topology. Following André [**Andr**] we denote the category of Nisnevich sheaves with transfers by[2] $\mathsf{Nis}_{\mathsf{tr}}(k)$.

Motivic Complexes. Let F be a presheaf of abelian groups on $\mathsf{Sm}(k)$. The *Suslin complex* $C_*(F)$ is the complex of presheaves on $\mathsf{Sm}(k)$ defined by

$$C_n(F)(U) = F(U \times \Delta^n)$$

whose differentials are alternating sums of pullbacks to the faces. By definition the sheaf $C_n(F)$ is placed in degree $-n$, making $C_*(F)$ a complex that is bounded above. If F is a Nisnevich sheaf with transfers, $C_*(F)$ is a bounded above complex of Nisnevich sheaves with transfers.

Let $D^-(\mathsf{Nis}_{\mathsf{tr}}(k))$ be the derived category of bounded above complexes of Nisnevich sheaves with transfers. The category $\mathsf{DM}^{\mathrm{eff}}_-(k)$ of *effective motivic complexes* is the full subcategory of $D^-(\mathsf{Nis}_{\mathsf{tr}}(k))$ of bounded above complexes of Nisnevich sheaves with transfers whose cohomology sheaves are homotopy invariant.

If F is a Nisnevich sheaf with transfers, the Suslin complex $C_*(F)$ has homotopy invariant cohomology sheaves. Hence we obtain a functor

$$C_* : \mathsf{Nis}_{\mathsf{tr}}(k) \to \mathsf{DM}^{\mathrm{eff}}_-(k).$$

This construction can be used to attach a motivic complex to *every* k-variety X as follows. We denote by $\mathbb{Z}_{\mathsf{tr}}(X)$ the presheaf with transfers defined by

$$\mathbb{Z}_{\mathsf{tr}}(X)(Y) = \mathsf{Corr}_{\mathrm{fin}}(Y, X), \quad Y \in \mathsf{Sm}(k).$$

One can show that $\mathbb{Z}_{\mathsf{tr}}(X)$ is a Nisnevich sheaf with transfers (cf. [**Maz-Vo-We**, Lemma 6.2]). Hence $\mathbb{Z}_{\mathsf{tr}}$ induces a functor

$$\mathbb{Z}_{\mathsf{tr}} : \mathbf{H}^b(\mathsf{SmCor}(k)) \to D^-(\mathsf{Nis}_{\mathsf{tr}}(k)).$$

The *motivic complex* of X is the class of

$$C_*(X) := C_*(\mathbb{Z}_{\mathsf{tr}}(X))$$

in $\mathsf{DM}^{\mathrm{eff}}_-(k)$. Concretely, $C_n(X)$ is the presheaf on $\mathsf{SmCor}(k)$ defined by

$$C_n(X)(U) = \mathsf{Corr}_{\mathrm{fin}}(U \times \Delta^n, X).$$

Note that if in particular also X is smooth, then $\mathbb{Z}_{\mathsf{tr}}(X)$ is the presheaf on $\mathsf{SmCor}(k)$ represented by X.

Voevodsky's main technical result is the following; see [**Friedl-Su-Vo**, Chapter 5, Prop. 3.2.3 and Thm. 3.2.6].

THEOREM (Voevodsky). *We have:*
(i) (Localization Theorem) The functor C_* extends to a functor

$$RC_* : D^-(\mathsf{Nis}_{\mathsf{tr}}(k)) \to \mathsf{DM}^{\mathrm{eff}}_-(k).$$

This functor identifies $\mathsf{DM}^{\mathrm{eff}}_-(k)$ *with the localization of* $D^-(\mathsf{Nis}_{\mathsf{tr}}(k))$ *with respect to the thick subcategory generated by complexes of the form*

$$\mathbb{Z}_{\mathsf{tr}}(X \times \mathbb{A}^1) \to \mathbb{Z}_{\mathsf{tr}}(X), \quad X \in \mathsf{Sm}(k).$$

(ii) (Embedding Theorem) The functor

$$RC_* \circ \mathbb{Z}_{\mathsf{tr}} : \mathbf{H}^b(\mathsf{SmCor}(k)) \to \mathsf{DM}^{\mathrm{eff}}_-(k)$$

---

[2]In [**Maz-Vo-We**] this category is denoted by $\mathsf{Sh}_{\mathrm{Nis}}(\mathrm{Cor}(k))$.

factors through a functor $i : \mathsf{DM}^{\mathrm{eff}}_{\mathrm{gm}}(k) \to \mathsf{DM}^{\mathrm{eff}}_{-}(k)$. *The functor* i *is a full embedding and satisfies*

$$i(M(X)) = RC_*(\mathbb{Z}_{\mathsf{tr}}(X)) \simeq C_*(X).$$

So in the end one obtains a commutative diagram

$$
\begin{array}{ccc}
\mathbf{H}^b(\mathsf{SmCor}(k)) & \xrightarrow{\;\mathbb{Z}_{\mathsf{tr}}\;} & D^-(\mathsf{Nis}_{\mathsf{tr}}(k)) \\
\downarrow{\scriptstyle M} & & \downarrow{\scriptstyle RC_*} \\
\mathsf{DM}^{\mathrm{eff}}_{\mathrm{gm}}(k) & \xrightarrow{\;i\;} & \mathsf{DM}^{\mathrm{eff}}_{-}(k).
\end{array}
$$

Construction of the Motives $M(X)$ and $M^c(X)$ for $X \in \mathsf{Var}(k)$. We have seen that the definitions of $\mathbb{Z}_{\mathsf{tr}}(X)$ and the motivic complex $C_*(X)$ make sense for every $X \in \mathsf{Var}(k)$. Hence one can define the motive without compact support $M(X)$ as the class of $C_*(X)$ in $\mathsf{DM}^{\mathrm{eff}}_{-}(k)$. By Voevodsky's embedding theorem this definition extends the definition of $M(X)$ for $X \in \mathsf{Sm}(k)$ given in section 9.3.1 to $\mathsf{Var}(k)$.

Given $X \in \mathsf{Sm}(k)$ and $Y \in \mathsf{Var}(k)$, the group $\mathsf{Corr}_{\mathrm{q-fin}}(X, Y)$ of *quasi-finite correspondences* from X to Y is the abelian group generated by integral subschemes $Z \subset X \times Y$ such that $p_X : Z \to X$ is quasi-finite and dominant over an irreducible component of X.

This enables us to define a Nisnevich sheaf with transfers $\mathbb{Z}^c_{\mathsf{tr}}(X)$ for every $X \in \mathsf{Var}(k)$ by

$$\mathbb{Z}^c_{\mathsf{tr}}(X)(Y) = \mathsf{Corr}_{\mathrm{q-fin}}(Y, X).$$

The *motivic complex with compact support* of X is

$$C^c_*(X) = C_*(\mathbb{Z}^c_{\mathsf{tr}}(X)).$$

The *motive with compact support* $M^c(X)$ is then defined as the class of $C^c_*(X)$ in $\mathsf{DM}^{\mathrm{eff}}_{-}(k)$.

One can show that for every $X \in \mathsf{Var}(k)$ the motives $M(X)$ and $M^c(X)$ belong to the triangulated subcategory $\mathsf{DM}^{\mathrm{eff}}_{\mathrm{gm}}(k)$ using the blow-up and Gysin triangles [**Friedl-Su-Vo**, Chapter 5, Corollaries 4.1.4 and 4.1.6].

Bibliography

Ab-K-M-W.	ABRAMOVICH, D., K. KARU, K. MATSUKI AND J., WŁODARCZYK: Torification and factorization of birational maps, J. Amer. Math. Soc. **15**, 531–572 (2002)
Andr.	ANDRÉ, Y.: *Une introduction aux motifs*, Panoramas et synthèses **17**, Soc. Math. de France, Paris, (2004)
Andr-Ka-O'S.	ANDRÉ, Y., B. KAHN AND P. O'SULLIVAN: Nilpotence, radicaux et structures monoïdales, Rend. Sem. Mat. Univ. Padova **108**, 107–291 (2002)
Ang-MüS98.	DEL ANGEL, P. L. AND S. MÜLLER-STACH: Motives of uniruled 3-folds, Compositio Math. **112**, 1–16 (1998)
Ang-MüS00.	DEL ANGEL, P. L. AND S. MÜLLER-STACH: On Chow motives of 3-folds, Trans. Amer. Math. Soc. **352**, 1623–1633 (2000)
Bal-Sc.	BALMER, P. AND M. SCHLICHTING: Idempotent completion of triangulated categories, J. Algebra **236**, 819–834 (2001)
BanRol.	DEL BAÑO ROLLIN, S.: On motives and moduli spaces of stable vector bundles over a curve, PhD thesis (1998). Available at `http://www.ma1.upc.edu/recerca/preprints/preprints-1998/Fitxers/` `9803bano.ps/at_download/file`
Bar-Hu-Pe-VdV.	BARTH, W., K. HULEK, C. PETERS AND A. VAN DE VEN: *Compact Complex Surfaces, Second Enlarged Edition*, Ergebnisse der Math., **3**, Springer Verlag, Berlin, etc. (2003)
Beau77.	BEAUVILLE, A.: Variétés de Prym et jacobiennes intermédiaires, Ann. Sci. École Norm. Sup. **10**, 309–391 (1977)
Beau83.	BEAUVILLE, A.: Quelques remarques sur la transformation de Fourier dans l'anneau de Chow d'une variété abélienne, in Lect. Notes in Math. **1016**, Springer Verlag, Berlin, etc., 238–260 (1983)
Beau86.	BEAUVILLE, A.: Sur l'anneau de Chow d'une variété abélienne, Math. Ann. **273**, 647–651 (1986)
Beil.	BEILINSON, A.: Higher regulators and values of L-functions, Journ. Soviet Math **80**, 2036–2070 (1985)
Beil-Ber-Del.	BEILINSON, A., J. BERNSTEIN AND P. DELIGNE: Faisceaux pervers, in *Analyse et topologie sur les espaces singuliers I*, Astérisque **100**, (1982)
Ber.	BERTHELOT, P.: *Cohomologie cristalline*, Lect. Notes in Math. **407** Springer Verlag, Berlin, etc. (1974).
Bitt.	BITTNER, F.: The universal Euler characteristic for varieties of characteristic zero, Compositio Math. **140**, 1011–1032 (2004)
Blo76.	BLOCH, S.: Some elementary theorems about algebraic cycles on Abelian varieties, Invent. Math. **37**, (1976), 215–228.
Blo79.	BLOCH, S.: Torsion algebraic cycles and a theorem of Roitman, Compositio Math. J. **39**, 107–127 (1979)
Blo80.	BLOCH, S: *Lectures on algebraic cycles*, Duke Univ. Press. **4**, (1980), second edition: New Math. Monographs **16**, Cambridge University Press (2010)
Blo-Sri.	BLOCH AND S, V. SRINIVAS: Remarks on correspondences and algebraic cycles, Amer. J. Math. **105**, 1235–1253 (1983)
Bo-Lu-Ra.	BOSCH, S., W. LÜTKEBOHMERT AND M. RAYNAUD: *Néron models*, Springer Verlag, Berlin etc. (1990)
Bourb.	BOURBAKI: *Algèbre, livre II, Chap. 8, Modules et anneaux semi-simples*, Hermann, Paris 1958.

Ca-Mi07.   DE CATALDO, M. AND MIGLIORINI, L.: Intersection Forms, Topology of Maps and Motivic Decomposition for Resolutions of Threefolds, in [**Nag-Pe**], (2007), 114–150

Ca-Mi09.   DE CATALDO, M. AND MIGLIORINI, L.: The decomposition theorem, perverse sheaves and the topology of algebraic maps, Bulletin of the Amer. Math. Soc. **46**, 535–633 (2009)

Chev58.   CHEVALLEY, C.: *Anneaux de Chow et Applications*, Sém. Chevalley (1958)

Chow54.   CHOW, W. L.: The Jacobian variety of an algebraic curve, Amer. J. Math. **74**, 895–909 (1954)

Chow55.   CHOW, W. L.: Abstract theory of the Picard and Albanese varieties, http://mathnt.mat.jhu.edu/mathnew/Chow/Articles/ch55c.pdf

Chow56.   CHOW, W. L.: On equivalence classes of cycles in an algebraic variety, Ann. of Math. **64**, 450–479 (1956)

Chow-vW.   CHOW, W. L. AND B. VAN DER WAERDEN: Zur algebraische Geometrie IX, Math. Ann. **113**, 692–704 (1937)

Clem.   CLEMENS, C. H.: Homological equivalence, modulo algebraic equivalence, is not finitely generated, Publ. Math. IHÉS **58**, 19–38 (1984).

Cor-Ha00.   CORTI, A AND M. HANAMURA: Motivic decomposition and intersection Chow groups I, Duke Math. Journal **103**, 459–522 (2000).

Cor-Ha07.   CORTI, A AND M. HANAMURA: Motivic decomposition and intersection Chow groups II, Pure Appl. Math. Q. **3**, 181–203 (2007)

Col-Se.   *Grothendieck-Serre correspondence.* Edited by Pierre Colmez and Jean-Pierre Serre, Documents Mathématiques (Paris) **2**, Soc.Math. de France, Paris, 2001.

Cu-Re.   CURTIS, C. W. AND I. REINER: *Representation theory of finite groups and associative algebras*, Wiley-Interscience, New York (1962)

DJ.   DE JONG, A. J.: A note on Weil cohomology (2007). Available at http://www.math.columbia.edu/~dejong/seminar/note_on_weil_cohomology .pdf

Del68.   DELIGNE, P.: Théorème de Lefschetz et critère de dégénérescence de suites spectrales, Publ. Math. IHÉS **35**, 107–126 (1968)

Del70.   DELIGNE, P.: Théorie de Hodge **I**, in *Actes du congrès international Nice 1970*, Gauthiers-Villars (1971) 425–430

Del71.   DELIGNE, P.: Théorie de Hodge II, Publ. Math. IHÉS **40**, 5–58 (1971)

Del74a.   DELIGNE, P.: Théorie de Hodge III, Publ. Math. IHÉS **44**, 5–77 (1974)

Del74b.   DELIGNE, P.: La conjecture de Weil. I, Publ. Math. IHÉS **43**, 273–307 (1974)

Del80.   DELIGNE, P.: La conjecture de Weil II, Publ. Math. IHÉS **52**, 137–252 (1980)

Del-Mi-Og-Sh.   DELIGNE, P., J. S. MILNE, A. OGUS AND K. SHIH: *Hodge cycles, motives and Shimura varieties*, Lect. Notes in Math. **900**, Springer Verlag, Berlin, etc. (1982)

Dem.   DEMAZURE, M. : Motifs des variétés algébriques, Sém. Bourbaki 365, Lect. Notes in Math. 180, Springer Verlag, Berlin, etc. (1971)

Den-Mu.   DENINGER, C. AND J. P. MURRE: Motivic decomposition of abelian schemes and the Fourier transform, Journal f. die reine und angew. Math. **422**, 201–219 (1991)

Elm-Ka-Me.   ELMAN, R., N. KARPENKO, AND A. MERKURJEV: *The algebraic and geometric theory of quadratic forms*, American Mathematical Society Colloquium Publications **56**, Amer. Math. Soc., Providence, RI (2008)

Friedl-La.   FRIEDLANDER, E. AND H. BLAINE LAWSON: Moving Algebraic Cycles of Bounded Degree, Invent. Math. **132**, 91–119 (1998)

Friedl-Su-Vo.   FRIEDLANDER, E., A. SUSLIN, AND V. VOEVODSKY: *Cycles, transfers, and motivic homology theories.* Ann. of Math. Studies **143**. Princeton University Press, Princeton, NJ, 2000

Ful.   FULTON, W.: *Intersection Theory*, Springer Verlag, New-York etc. (1984)

Ful-Ha.   FULTON, W. AND J. HARRIS: *Representation theory. A first course.* Grad. Texts in Math. **129**. Springer-Verlag, New York, 1991.

Gil-So.   GILLET, H. AND C. SOULÉ: Descent, motives and K-theory. Gillet, Journal f. die reine u. angew. Math. **478**, 127–176 (1996)

Gor-Ha-Mu.   GORDON, B. BRENT, M. HANAMURA AND J.P. MURRE: Relative Chow-Künneth projectors for modular varieties, J. f. die reine u. angew. Math. **558**, 1-14 (2003).

Griff-Ha.   GRIFFITHS, PH. AND J. HARRIS: *Principles of algebraic geometry*, Wiley, New York (1978)

Griff69.   GRIFFITHS, P.: On the periods of certain rational integrals I, II, Ann. Math. **90**, 460–495, 498–541, respectively (1969)

Groth58.   GROTHENDIECK, A.: Sur une note de Mattuck-Tate. J. f. die reine u. angew. Math. **200**, 208–215 (1958)

Groth62.   GROTHENDIECK, A.: *Fondements de la géométrie algébrique*, Sém. Bourbaki 1957–1962, Secr. Math., Paris (1962)

Groth69a.   GROTHENDIECK, A.: Standard conjectures on algebraic cycles, in *Algebr. Geom. Bombay Colloquium 1968*, Oxford, 193–199 (1969)

Groth69b.   GROTHENDIECK, A.: *Hodge's general conjecture is false for general reasons*, Topology **8**, 299–303 (1969)

Groth85.   GROTHENDIECK, A.: *Récoltes et semailles: réflexions et témoignages sur un passé de mathématicien*, Montpellier, 1985 (unpublished)

Gu-Na.   GUILLEN, F. AND V. NAVARRO AZNAR: Un critère d'extension d'un foncteur défini sur les schémas lisses, Publ. Math. IHÉS **95**, 1–91 (2002)

Hana95.   HANAMURA, M.: Mixed motives and algebraic cycles I, Math. Res. Lett. **2**, 811–821 (1995)

Hana04.   HANAMURA, M.: Mixed motives and algebraic cycles II, Invent. Math. **158**, 105-179 (2004)

Harts.   HARTSHORNE, R.: *Algebraic Geometry*, Springer Verlag, Berlin Heidelberg etc., Grad. Texts in Math. **52** (1977)

Hub-MüS.   HUBER, A. AND S. MÜLLER-STACH: On the relation between Nori Motives and Kontsevich Periods, preprint `arXiv:1105.0865`

Ill.   ILLUSIE, L.: Report on crystalline cohomology, in *Algebraic geometry, Arcata 1974*, Proc. of Symp. Amer. Math. Soc. **29**, 459–478 (1975)

Jann92.   JANNSEN, U.: Motives, numerical equivalence and semi-simplicity, Invent. Math. **107**, 447–452 (1992)

Jann94.   JANNSEN, U.: Motivic sheaves and filtration on Chow groups, in [**Jann-Kl-Se**], 245–302 (1994

Jann03.   JANNSEN, U.: Algebraic cycles and motives, Course notes, Tokyo Univ. 2003/2004 (unpublished).

Jann07.   JANNSEN, U.: On finite dimensional motives and Murre's conjecture, in [**Nag-Pe**], 112–142 (2007)

Jann-Kl-Se.   *Motives. Proceedings of the AMS-IMS-SIAM Joint Summer Research Conference held at the University of Washington, Seattle, Washington, July 20–August 2, 1991* (Ed. by Uwe Jannsen, Steven Kleiman and Jean-Pierre Serre) Proc. Symp. Pure Math. **55**, Part I and II, Amer. Math. Soc., Providence, RI, (1994)

Kahn-M-P.   KAHN, B., J. P. MURRE AND C. PEDRINI: On the Transcendental Part of the Motive of a Surface, in [**Nag-Pe**], 2007

Katz-Me.   KATZ, N. AND W. MESSING: Some consequences of the Riemann hypothesis for varieties over finite fields, Invent. Math. **23**, 73–77 (1974)

Kim.   KIMURA, K.: Murre's conjectures for certain product varieties, J. Math. Kyoto Univ. **47**, 621-629 (2007)

Kimu.   KIMURA, S-I: Chow groups are finite dimensional, in some sense, Math. Ann. **331**, 173–201 (2005)

Kimu-Mur.   KIMURA, S-I AND J. P. MURRE: On natural isomorphisms of finite dimensional motives and applications to Picard motives, in: *Proceedings Int. Coll. "Cycles, Motives and Shimura varieties"*, V. Srinivas (Ed.), Mumbai 2008, Naroshi Publ. House (New Delhi) and AMS (2010)

Kimu-Vi.   KIMURA, S.-I. AND A. VISTOLI: Chow rings of infinite symmetric products, Duke Math. J. **85**, 411–430 (1996)

Klei68.   KLEIMAN, S. L.: Algebraic cycles and the Weil conjectures, in *Dix exposés sur la cohomologie des schémas*, North Holland, 359–386 (1968)

Klei70.   KLEIMAN, S. L.: Motives in Algebraic Geometry, in *Conf. Oslo*, 53–82 (1970)

Klei94.   KLEIMAN, S. L.: The standard conjectures, in [**Jann-Kl-Se**] 3–20 (1994)

Klei05.　　　KLEIMAN, S. L. : The Picard Scheme, Part 5, in *Fundamental Algebraic Geometry; Grothendiecks FGA explained*, Math. Surveys and Monographs **123** Am. Math. Soc., Providence R.I. (2005)

Kü93.　　　KÜNNEMANN, K.: A Lefschetz decomposition for Chow motives of abelian schemes, Invent. Math. **113**, 85–102 (1993)

La.　　　LANG, S.: *Abelian varieties*, Interscience, New York (1959) and Springer Verlag, Berlin, etc.(1983)

Le-C.　　　LE, J. AND CHEN, X.-W.: *Karoubianness of a triangulated category*, J. of Algebra **310**, 452-457 (2007)

Lev98.　　　LEVINE, M.: *Mixed Motives* Math. Surveys and Monographs, Vol. **57**, Am. Math. Soc., Providence R.I. (1998)

Lev05.　　　LEVINE, M.: Mixed motives, in: *Handbook of K-theory.* Vol. **1**, **2**, 429-521, Springer Verlag, Berlin, etc. (2005)

Lev08.　　　LEVINE, M.: Six Lectures on Motives, in: Some recent developments in algebraic K-theory, ICTP Lecture Notes **23**, 131-227 (2008)

Liebm.　　　LIEBERMAN, D.: Numerical and homological equivalence of algebraic cycles on Hodge manifolds, Amer. J. Math. **90**, 366–374 (1968)

Manin.　　　MANIN, YU.I.: Correspondences, motives and monoidal transformations, Math. U.S.S,R. Sbornik **6**, 439–470 (1968)

Mats52.　　　MATSUSAKA, T. : On the algebraic construction of the Picard variety I, Jap. J. Math. **21** 217–235 (1952)

Mats57.　　　MATSUSAKA, T.: The criteria for algebraic equivalence and the torsion group, Amer. J. Math. **79** 53–66 (1957)

May.　　　MAY, P.: *Simplicial objects in algebraic topology*, Van Nostrand (1967)

Mazur.　　　MAZUR, B.: What is ... a motive? Notices Amer. Math. Soc. **51**, 1214–1216 (2004)

Maz-Vo-We.　　　MAZZA, C, V. VOEVODSKY AND C. WEIBEL: *Lecture Notes on Motivic Cohomology* CMI/AMS, Claymath Publ. **7** (2006)

Mil80.　　　MILNE, J. S.: *Etale Cohomology*, Princeton Univ. Press, Princeton (1980)

Mil82.　　　MILNE, J. S.: Zero cycles on algebraic varieties in nonzero characteristic: Rojtman's theorem, Compositio Math. **47**, 271–287 (1982)

Mil98.　　　MILNE, J. S.: Lectures on Etale Cohomology. Latest version (2012) available at `http://www.jmilne.org/math/CourseNotes/lec.html` (1998)

Mil09.　　　MILNE, J. S.: Motives – Grothendieck's dream. Latest version (2012) at `http://jmilne.org/math/xnotes/MOT.pdf` (2009)

Muk.　　　MUKAI, SH.: Duality between $D(X)$ and $D(\hat{X})$ with its application to Picard sheaves, Nagoya Math. J. **81**, 153–175 (1981)

MüS-Pe.　　　*Transcendental aspects of algebraic cycles . Proceedings of the Summer School held in Grenoble, June 18–July 6, 2001.* Edited by S. Müller-Stach and C. Peters. London Mathematical Society Lecture Note Series, **313**, Cambridge University Press, Cambridge (2004)

MüS-Sa.　　　MÜLLER-STACH, S. AND M. SAITO: Relative Chow-Kuenneth decompositions for morphisms of threefolds `http://arxiv.org/abs/0911.3639`, to appear in J. f. die reine u. angew. Math.

Mum69.　　　MUMFORD, D.: Rational equivalence of zero-cycles on surfaces, J. Math. Kyoto Univ. **9**, 195–204 (1969)

Mum74.　　　MUMFORD, D.: *Abelian varieties* (with appendices by C.P. Ramanujam and Yuri Manin) [Second edition] Oxford Univ. Press, for the Tata Institute of Fundamental Research, Bombay (1974)

Mur90.　　　MURRE, J. P. : On the motive of an algebraic surface, Journal f. die reine u. angew. Math. **409**, 190–204 (1990)

Mur93.　　　MURRE, J. P.: On a conjectural filtration on the Chow groups of an algebraic variety I and II, Indag. Mathem. **4**, 177–188 and 189–201(1993)

Nag-Pe.　　　NAGEL, J. AND C. PETERS: *Algebraic Cycles and Motives*, Proceedings of the Annual EAGER Conference And Workshop On the Occasion of J. Murre's 75–th Birthday August 30– September 3, 2004, Leiden University, the Netherlands, Lond. Math. Soc. Lect. Notes Series, Cambridge University Press **343** & **344** (2007)

Nag-Sa.   NAGEL, J, AND SAITO, M.: Relative Chow-Kunneth decompositions for conic bundles and Prym varieties, Int. Math. Res. Not. **16**, 2978–3001(2009)

Ner.   NÉRON, A.: La théorie de la base pour les diviseurs sur les variétés algébriques, in *Deuxième Colloque de Géométrie Algébrique, Liège, 1952* 119–126, Georges Thone, Liège; Masson & Cie, Paris, (1952)

Nori.   M. NORI: *Lectures at TIFR*, unpublished 32 pages.

Pe-St.   PETERS, C. AND J. STEENBRINK: *Mixed Hodge Theory*, Ergebnisse der Math, Wiss. **52**, Springer Verlag, Berlin, etc. (2008)

Pie.   PIERCE, R.S: *Associative Algebras*, Grad. Text in Math. **88**, Springer Verlag, Berlin, etc. (1982)

Rob.   ROBERTS, J.: Chow's moving lemma, Appendix 2 to: *"Motives" (Algebraic geometry, Oslo 1970 (Proc. Fifth Nordic Summer School in Math.)*, pp. 53–82, Wolters-Noordhoff, Groningen, (1972)

Roi.   ROJTMAN, A.A.: The torsion of the group of 0-cycles modulo rational equivalence, Ann. of Math. (2) **111** 553–569, (1980)

Sam55.   SAMUEL, P.: *Méthodes d'Algèbre abstraite en Géométrie Algébrique*, Erg. Math. **4**, Springer Verlag, Berlin etc. (1955)

Sam56.   SAMUEL, P.: Rational equivalence of arbitrary cycles, Amer. J. Math. **78**, 383–400 (1956)

Sam58.   SAMUEL, P.: Relations d'équivalence en géométrie algébrique, in *Proc. I.C.M. 1958, Edinburgh*, Cambridge Univ. Press, Cambridge (1960)

Scholl.   SCHOLL, A. J. : Classical Motives, in [**Jann-Kl-Se**] 163–187 (1994)

Schwar.   SCHWARZENBERGER, R.: Jacobians and symmetric products, Illinois J. Math. **7** 257–268 (1963)

Segre.   SEGRE, B.: Intorno ad un teorema di Hodge sulla teoria della base per le curve di una superficie algebrica, Ann. Mat. Pura Appl. **16**, 157-163 (1937).

Serm.   ŠERMENEV, A. M.: The motif of an Abelian variety (English. Russian original), Funct. Anal. Appl. **8** 47–53 (1974); translation from Funkts. Anal. Prilozh. **8**, No.1, 55–61 (1974)

Serre56.   SERRE, J.-P.: Géometrie algébrique et géométrie analytique, Ann. Inst. Fourier **6**, 1–42 (1956)

Serre57.   SERRE, J.-P.: *Algèbre locale, Multiplicités*, Springer Lecture Notes in Math, **11** Second edition Springer Verlag, Berlin, etc.(1965)

Serre77.   SERRE, J.-P.: *Linear representations of finite groups* (transl. from the French by Leonard L. Scott), Springer Verlag, Berlin etc. (1977)

Serre91.   SERRE, J.-P.: Motifs, in Astérisque **198-200**, 333–349 (1991)

Sesh62.   SESHADRI, C. S.: Variété de Picard d'une variété complète, Ann. Mat. Pura Appl. (4) **57**, 117–142 (1962)

Sev.   SEVERI, F.: *Serie, Sistemi d'Equivalenza e Corrispondenze Algebriche sulle Varietà Algebriche, Vol. 1* Edizioni Cremonense, Rome (1942)

SGA4.   *Théorie des topos et cohomologie étale des schémas*, 1963–1964 (Topos theory and etale cohomology), Lect. Notes in Math, **269**, **270** and **305**, Springer Verlag, Berlin etc. (1972/3)

Verd.   VERDIER, J.-L.: *Des Catégories Dérivées des Catégories Abéliennes* Astérisque **239** Soc. Math. de France, Paris; Amer. Math. Soc., Providence R.I. (1996)

Voe95.   VOEVODSKY, V. : A nilpotence theorem for cycles algebraically equivalent to zero, Int. Math. Res.. Notes **4**, 1–12 (1995)

Voe00.   VOEVODSKY, V. : Triangulated categories of motives over a field, in *Cycles, transfers and motivic cohomology theories (by V. Voevodsky, A. Suslin and E.M. Friedlander)*, Princeton Univ. Press, Princeton (2000)

Voi92.   VOISIN, CL.: Une approche infinitésimale du théorème de H. Clemens sur les cycles d'une quintique générale de $\mathbb{P}^4$, J. Algebraic Geom. **1**, 157–174 (1992)

Voi96.   VOISIN, CL.: Remarks on zero-cycles of self-products of varieties, in: *Moduli of vector bundles, Proc. of the Taniguchi symposium on vector bundles, December 1994*, ed. by M. Maruyama, pp. 265–285. Decker (1996)

Voi03.   VOISIN, CL.: *Hodge Theory and Complex Algebraic Geometry II*, Cambridge Studies in Adv. Math. **77** Cambridge Univ. Press, Cambridge (2003)

Weib.   WEIBEL, C.: *An introduction to homological algebra*, Cambridge Studies in Adv. Math. **38**, Cambridge Univ. Press, Cambridge (1994)

Weil46.   WEIL, A.: *Foundations of algebraic geometry* 1962 Colloquium publications, Amer. Math. Soc., Providence R.I. **29**, (1946) (revised 1984)

Weil48.   WEIL, A.: *Variétés abéliennes et courbes algébriques*, Hermann, Paris, (1948)

Weil54.   WEIL, A.: Sur les critères d'équivalence en géométrie algébrique, Math. Ann. **128**, 95–127 (1954), also in *Collected papers, II*, Springer Verlag, Berlin, etc. (1980) (second printing)

Zar.   ZARISKI, O.: *Algebraic Surfaces*; with appendices by S.S. Abhyankar, J. Lipman, and D. Mumford, 2nd supplemented edition, Springer Verlag, Berlin, etc. (1971)

Index of Notation

$\mathrm{DM}^{\mathrm{gm}}_{\mathrm{eff}}(k)$: effective geometric motives over k, 132

e_λ: element in $R(\mathfrak{S}_n)$ defined by partition λ, 44

e_{alt}: the n-the alternator in $R(\mathfrak{S}_n)$, 44

e_{sym}: the n-th symmetrizer in $R(\mathfrak{S}_n)$, 44

$F^\bullet_{BB}$: Bloch-Beilinson filtration, 85

f^*: (flat) pull back of cycles, 2

f_*: push forward of cycles, 2

$G(X)$: Gysin complex of X, 127

GrVect_F: category of finite dimensional graded F-vector spaces, 7

$\mathrm{Griff}^i(X)$: i–th Grifiths group of X, 18

$H^2_{\mathrm{trans}}(X)$: transcendental 2–cohomology for surface X, 17

$H^i_{\mathrm{crys}}(X/W(k))$: i–th crystalline cohomology of X, 8

$H^i_{\mathrm{dR}}(X)$: i–th (algebraic) de Rham cohomology of X, 8

$H^i_{\mathrm{B}}(X)$: i–th Betti cohomology of X, 8

$H^i_{dR}(X_{\mathrm{an}};\mathbb{C})$: i–th (analytic) de Rham cohomology of X, 8

$H^i_{\mathrm{\acute{e}t}}(X_{\bar{k}},\mathbb{Q}_\ell)$: i–th étale cohomology of X, 8

$H^p(X,\mathbb{Z}(q))$: (p,q)–th motivic cohomology group of X, 133

$h_\sim$: functor of motives of smooth varieties with respect to an equivalence $\sim$, 26

$i(V \cdot W; Z)$: intersection number of V and W along Z, 2

$\mathrm{IC}^\bullet_S$: intersection complex of S, 109

$\mathrm{IH}^*(S)$: intersection cohomology groups of S, 109

$\mathrm{J}(C)$: jacobian of the curve C, 15

$K_0\mathrm{Mot}_\sim(k)$: K-group for motives, 125

$K_0\mathrm{Var}(k)$: K-group of k-varieties, 124

$\mathbf{L}_S$: Lefschetz motive over S, 107

$\mathbf{L}_\sim$: Lefschetz motive, 28

$M(X)$: Voevodsky-motive of X, 132

$M^c(X)$: Voevodsky-motive of X with compact support, 133

$\mathrm{Mot}^{\mathrm{eff}}_\sim(k)$: the category of effective motives over k, 25

$\mathrm{Mot}_\sim(k)$: pure motives over k, 25

NM: category of Grothendieck motives, 26

$\mathrm{NS}(X)$: Néron-Severi group of X, 13

ord: the order homomorphism, 4

$\mathrm{Perv}(S)$: perverse sheaves on S, 111

$\mathrm{Pic}^0_{\mathrm{red}}(X)$: Picard variety of X, 14

$R(G)$: group ring on G with $\mathbb{Q}$-coefficients, 44

$\mathrm{SmCor}(k)$: category of smooth correspondences over k, 131

$\mathrm{SmProj}(k)$: the category of smooth projective varieties over k, 1

$\mathrm{Sym}^n(M)$: n-th symmetric motive on M, 45

$T(X)$: Albanese kernel for X , 17

$T_\ell(A)$: ℓ–adic Tate group of A, 73

$\mathbb{T}_\lambda M$: motive built from M and partition λ, 45

$\mathbf{T}_\sim$: Tate motive, 28

$V \cdot W$: intersection cycle of V and W, 2

$\mathrm{Var}(k)$: category of k-varieties, 124

$\mathcal{V}(S)$: category of varieties smooth and projective over a quasi-projective smooth base, 113

$W(X)$: weight complex of X, 127

$X \vdash Y$: correspondence from X to Y, 23

$^T Z$: transpose of the correspondence Z, 3

$\mathbf{Z}^i_\otimes(X)$: codim-i cycles smash nilpotently equivalent to 0, 7

$\mathbf{Z}^i(X)_F$: codim-i cycles on X with F-coefficients, 1

$\mathbf{Z}^i_{\mathrm{hom}}(X)$: codim–$i$ cycles homologically equivalent to zero, 9

$\mathbf{Z}^i_{\mathrm{num}}(X)$ codim–i cycles numerically equivalent to zero, 10

$\mathbf{Z}^i(X)$: codim-i cycles on X, 1

$\mathbf{Z}^i_{\mathrm{alg}}(X)$: codim-$i$ cycles algebraically equivalent to zero, 6

$\mathbf{Z}^i_\sim(X)$: codim-i cycles on X equivalent to zero, 3

$\mathbf{Z}^i_{\mathrm{rat}}(X)$: codim-$i$ cycles on X modulo rational equivalence, 4

$\mathbf{Z}_\sim(X)_F$: codim-i cycles on X with F-coefficients equivalent to zero, 3

Index